Advances in Intelligent Systems and Computing

Volume 807

Series editor

Janusz Kacprzyk, Systems Research Institute, Polish Academy of Sciences, Warsaw, Poland
e-mail: kacprzyk@ibspan.waw.pl

The series "Advances in Intelligent Systems and Computing" contains publications on theory, applications, and design methods of Intelligent Systems and Intelligent Computing. Virtually all disciplines such as engineering, natural sciences, computer and information science, ICT, economics, business, e-commerce, environment, healthcare, life science are covered. The list of topics spans all the areas of modern intelligent systems and computing such as: computational intelligence, soft computing including neural networks, fuzzy systems, evolutionary computing and the fusion of these paradigms, social intelligence, ambient intelligence, computational neuroscience, artificial life, virtual worlds and society, cognitive science and systems, Perception and Vision, DNA and immune based systems, self-organizing and adaptive systems, e-Learning and teaching, human-centered and human-centric computing, recommender systems, intelligent control, robotics and mechatronics including human-machine teaming, knowledge-based paradigms, learning paradigms, machine ethics, intelligent data analysis, knowledge management, intelligent agents, intelligent decision making and support, intelligent network security, trust management, interactive entertainment, Web intelligence and multimedia.

The publications within "Advances in Intelligent Systems and Computing" are primarily proceedings of important conferences, symposia and congresses. They cover significant recent developments in the field, both of a foundational and applicable character. An important characteristic feature of the series is the short publication time and world-wide distribution. This permits a rapid and broad dissemination of research results.

More information about this series at http://www.springer.com/series/11156

Thanaruk Theeramunkong
Rachada Kongkachandra
Mahasak Ketcham · Narit Hnoohom
Pokpong Songmuang · Thepchai Supnithi
Kiyota Hashimoto
Editors

Advances in Intelligent Informatics, Smart Technology and Natural Language Processing

Selected Revised Papers from the Joint International Symposium on Artificial Intelligence and Natural Language Processing (iSAI-NLP 2017)

 Springer

Editors
Thanaruk Theeramunkong
Sirindhorn International Institute of
Technology (SIIT)
Thammasat University
Bangkadi Muang Pathumthani, Thailand

Rachada Kongkachandra
Faculty of Science and Technology,
Department of Computer Science
Thammasat University
Pathumthani, Thailand

Mahasak Ketcham
Department of Information Technology
Management, Faculty of Information
Technology
King Mongkut's University of Technology
North Bangkok (KMUTNB)
Bangkok, Thailand

Narit Hnoohom
Department of Computer Engineering,
Faculty of Engineering
Mahidol University
Nakorn Pathom, Thailand

Pokpong Songmuang
Faculty of Science and Technology,
Department of Computer Science
Thammasat University
Pathum Thani, Thailand

Thepchai Supnithi
National Science and Technology
Development Agency
Language and Semantic Technology
Laboratory, National Electronics and
Computer Technology Center
Pathum Thani, Thailand

Kiyota Hashimoto
Prince of Songkla University
Phuket, Thailand

ISSN 2194-5357 ISSN 2194-5365 (electronic)
Advances in Intelligent Systems and Computing
ISBN 978-3-319-94702-0 ISBN 978-3-319-94703-7 (eBook)
https://doi.org/10.1007/978-3-319-94703-7

Library of Congress Control Number: 2018958463

This Springer imprint is published by the registered company Springer Nature Switzerland AG
The registered company address is: Gewerbestrasse 11, 6330 Cham, Switzerland

Preface

The Twelfth Series of Symposium on Natural Language Processing (SNLP2017), which biannually held since 1993, has broadened to cover the knowledge in artificial Intelligence together with the knowledge in natural language processing. The new name becomes The Joint International Symposium on Artificial Intelligence and Natural Language Processing (iSAI-NLP 2017).

The iSAI-NLP 2017 is the collaboration of a number of universities in Thailand with the purpose to promote researches in artificial intelligence, natural language processing, and other related fields. As a unique, premier meeting of researchers, professionals, and practitioners, iSAI-NLP 2017 provides a place and an opportunity for them to discuss various current and advanced issues of interests in such areas. The iSAI-NLP 2017 is hosted by College of Information and Communication Technology, Rangsit University, and the Project of Administrative Corporation in Media Arts and Media Technology Curriculum, King Mongkut's University of Technology Thonburi, with the great support from Artificial Intelligence Association of Thailand (AIAT).

We accepted three special session proposals with the goal of exploring focused issues across various themes. Each paper in these proceedings was peer-reviewed by international reviewers in their respective areas to ensure the highest quality work. Forty-one papers are accepted to present in the conference. After an intense discussion during the conference, we selected 19 high-quality papers (38% of the 41 submitted papers) for this volume.

The volume contains the 19 papers were presented at the Joint International Symposium on Artificial Intelligence and Natural Language Processing Conference (August 27–29, 2017, in Prachuap Khiri Khan, Thailand, http://www.isai-nlp.org) and the three papers were presented at the Tenth International Conference on Knowledge, Information and Creativity Support Systems (November 12–14, 2015, in Phuket, Thailand).

Finally, we would like to thank the iSAI-NLP 2017 Executive Committees and Program Co-Chairs for entrusting us with the important task of chairing the main track and special session program, thus giving us an opportunity to grow through valuable academic learning experiences. We also would like to thanks all Co-Chairs for their tremendous and excellent work.

August 2017   Thanaruk Theeramunkong
                                                                          Thepchai Supnithi
                                                                   Rachada Kongkachandra
                                                                        Mahasak Ketcham
                                                                           Narit Hnoohom
                                                                       Kiyota Hashimoto
                                                                     Pokpong Songmuang

Organization

Organizing Committee

Honorary Co-chairs

Kulthorn Kasemsan	Rangsit University, Thailand
Paiboon Kiattikomol	King Mongkut's University of Technology Thonburi, Thailand
Vilas Wuwongse	Mahidol University, Thailand

General Co-chairs

Bulent Topcuoglu	Akdeniz University, Antalya, Turkey
Pranesh Kumar	University of Northern British Columbia, Canada
Thanaruk Theeramunkong	Sirindhorn International Institute of Technology, Thammasat University, Thailand
Thepchai Supnithi	National Electronics and Computer Technology Center, Thailand

Technical Program Co-chairs

Abid-UlRazzaq 1 Abu-Tair	British University, UAE
Kai Way Li	Chung Hua University, Taiwan
Kazuaki Maeda	Chubu University, Japan
Yoshinori Sagisaka	Waseda University, Japan
Kiyota Hashimoto	Prince of Songkla University, Thailand

Kritsada Sriphaew   Rangsit University, Thailand
Pokpong Songmuang   Thammasat University, Thailand
Thaweesak Yingthawornsuk   King Mongkut's University of Technology
                                     Thonburi, Thailand

Special Session Co-chairs

Chai Wutiwiwatchai   National Electronics and Computer Technology
                                     Center, Thailand
Sanparith Marukatat   National Electronics and Computer Technology
                                     Center, Thailand

Publicity Co-chairs

Kobkrit Viriyayudhakorn   Sirindhorn International Institute of Technology,
                                     Thammasat University, Thailand
Patiyuth Pramkeaw   King Mongkut's University of Technology
                                     Thonburi, Thailand

Publication Co-chairs

Chuleerat Jaruskulchai   Kasetsart University, Thailand
Mahasak Ketcham   King Mongkut's University of Technology
                                     North Bangkok, Thailand
Narit Hnoohom   Mahidol University, Thailand
Rachada Kongkachandra   Thammasat University, Thailand

Organizing Chair

Chutima Beokhaimook   Rangsit University, Thailand

Special Session Organizers

Rise of Innovative Applications of Artificial Intelligence Conference (RIAA)

Sakorn Mekruksavanich   University of Phayao, Thailand
Narit Hnoohom   Mahidol University, Thailand

Anuchit Jitpattanakul	King Mongkut's University of Technology North Bangkok, Thailand
Mahasak Ketcham	King Mongkut's University of Technology North Bangkok, Thailand
Kirk Scott	University of Alaska Anchorage, USA

Business Intelligence System (OBIS)

Kritchana Wongrat	Phetchaburi Rajabhat University, Thailand
Wiwit Suksangaram	Phetchaburi Rajabhat University, Thailand

Information Security and Privacy (ISA)

Ngoc Hong Tran	Vietnamese German University, Viet Nam
Leila Bahri	Koç University, Turkey

Special Session Committee

Chai Wutiwiwatchai	National Electronics and Computer Technology Center, Thailand
Sanparith Marukatat	National Electronics and Computer Technology Center, Thailand
Marut Buranarach	National Electronics and Computer Technology Center, Thailand
Choochart Haruechaiyasak	National Electronics and Computer Technology Center, Thailand
Thatsanee Chareonporn	Burapha University, Thailand

Program Committee

Ahmad Alsahaf	University of Groningen, Netherlands
Amnat Sawatnatee	Chandrakasem Rajabhat University, Thailand
Chenyu Shi	University of Groningen, Netherlands
Edison Muzenda	University of Science and Technology, South Africa
Estefania Talavera	University of Groningen, Netherlands
Hiroya Takamura	Tokyo Institute of Technology, Japan
Jiapan Guo	University of Groningen, Netherlands
Kei Eguchi	Fukuoka Institute of Technology, Japan
Mahasak Ketcham	King Mongkut's University of Technology North Bangkok, Thailand
Michael Pecht	University of Maryland, USA

Michel Plaisent   University of Quebec in Montreal, Canada
Narit Hnoohom   Mahidol University, Thailand
Narumol Chumuang   Muban Chombueng Rajabhat University,
                                            Thailand
Ngoc Hong Tran   Vietnamese German University, Viet Nam
Nicola Strisciuglio   University of Groningen, Netherlands
Panana Tangwannawit   Phetchabun Rajabhat University, Thailand
Patiyuth Pramkeaw   King Mongkut's University of Technology
                                            Thonburi, Thailand
Rachada Kongkachandra   Thammasat University, Thailand
Sanon Chimmanee   Rangsit University, Thailand
Siriporn Supratid   Rangsit University, Thailand
Sven Wohlgemuth   Center for Advanced Security Research
                                            Darmstadt, Germany
Thaweesak Yingthawornsuk   King Mongkut's University of Technology
                                            Thonburi, Thailand
Thittaporn Ganokratanaa   Chulalongkorn University, Thailand

Contents

Information Security and Privacy (ISA)

Knowledge, Information and Creativity Support Systems (KICSS)

iSAI-NLP

Building a Multimodal Language Resource for Thai Cuisine

Dhanon Leenoi[(⊠)], Wasan Na Chai, Witchawoan Mankhong,
and Thepchai Supnithi

Language and Semantic Technology Laboratory, National Electronics and
Computer Technology Center, Pathumthanee, Thailand
{dhanon.leenoi,wasan.na_chai,
thepchai.sup}@nectec.or.th, witchaworn.mk@gmail.com

Abstract. This paper presents the current status of the first standardized multimodal language resource for Thai cuisine. Scrupulously translated from Thai to other seven foreign languages: English and Italian; Spanish and Chinese; German and Japanese, with French, this contains 1,654 parallel lexical entries together with the illustrations and recorded voices. All nodes were attached and retained in the designed tree structure with fourteen categories and nineteen subcategories. To utilize a benefit of the project is to develop an application manifesting texts, images, speeches, and suggesting the dietary pattern of Thai food tailored to the personal health conditions and preferences.

Keywords: Thai cuisine · Multimodal · Language resource · Lexicography
Knowledge-based approach · Taxonomy · Machine learning

1 Introduction

Tout à fait, Thai cuisine becomes the world-famous for its succulent coloring and unique taste [1–3]. The renowned assemblage of sweet and sour; salty and spicy makes Thai food distinctive. Thai restaurants, nowadays, can be found in the world-major-cities. Certainly, in the menu, Thai dish names and ingredients have to be translated into foreign languages.

However, little has been done with the standardized Thai food terms, some relatively erroneous and unsystematic translations exist, since the classical lexicography, the dictionary-making, is relatively time-consuming and labor-intensive. Thus, Chalermprakiat Center of Translation and Interpretation – CCTI, Faculty of Arts, Chulalongkorn University collaborating with the National Electronic and Computer Technology Center – NECTEC would like to invent the gold- standard Thai food terms containing texts, speeches and images through a knowledge-based approach [4].

This study is of relevance because knowing how the machine links and learns simultaneously [5–7] on texts, images and speeches, then figures out how to perform important tasks by generalizing from examples [8] can be utilized in smart search, recommender systems, deceit detection, drug design, and many other applications.

This paper seeks to explain the crafting of a standardized multimodal language resource, Thai and English, Japanese and German, French and Chinese, Spanish and

© Springer Nature Switzerland AG 2019
T. Theeramunkong et al. (Eds.): iSAI-NLP 2017, AISC 807, pp. 3–13, 2019.
https://doi.org/10.1007/978-3-319-94703-7_1

Italian, in the sphere of Thai cuisine. This paper is divided into four main sections. Firstly, we propose the construction of multimodal language resource, followed by the methodology, then scenario setting with discussion, and conclude and envisage the future works finally.

2   Building a Multimodal Language Resource

Through five steps: (1) designing taxonomy and assembling wordlist, (2) composing definitions, (3) distributing and translating, (4) collecting and verifying, and (5) recording native speaker's voice and gathering illustrations, shown in Fig. 1, we propose the process of building a multimodal language resource for Thai cuisine together with caveats.

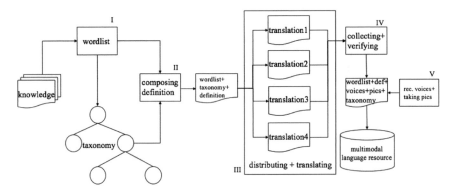

Fig. 1. Building a multimodal language resource

2.1   Designing Taxonomy and Assembling Wordlist

Thai word entries were manually collected from the authentic cookbooks and recipes, the local restaurants including fresh markets, and were classified into the two-level taxonomy which enumerated in fourteen categories (Table 1): (1) herbs and vegetables, (2) edible flowers, (3) fruits, (4) spices, (5) ingredients, seasoning and oils, (6) rice and cereal, (7) noodles, (8) meats, (9) seafood, (10) cooking verbs, (11) dish names, (12) beverages, (13) kitchen utensils, and (14) general culinary terms.

As shown in Table 1, containing in fourteen categories and nineteen sub-categories, 1,654 parallel word-entries were obtained; subsequently, the definitions had to be defined.

Table 1. Fourteen categories and whose sub-categories.

Categories	Subcategories	Entries
Herbs and vegetables		206
Edible flowers		24
Fruits		95
Spices		19
Ingredients, seasoning and oils		81
Rice and cereal		43
Noodles		19
Meats		106
Seafood		140
Cooking verbs		145
Dish names		395
	Snacks, appetizers, pre-meal dishes	(40)
	Salad, spicy salad	(25)
	Mild soup, clear soup, spicy soup	(32)
	Curry, Panaeng Curry	(31)
	Stir-fried and sautéed Dishes	(32)
	Deep-fried dishes	(11)
	Steamed and stewed dishes	(15)
	Casserole dishes	(4)
	Grilled and barbecued dishes	(10)
	One-plate dishes, individual rice dishes, noodles	((67)
	Chili dip, Nam Prik, chili relish	(17)
	Sauce, relish	(6)
	Egg and egg dishes	(30)
	Side dish, complimentary items, vegetable fritters	(6)
	Desserts	(69)
Beverages		81
Kitchen utensils		136
General culinary terms		164
	Flavors and textures	(30)
	Aromas, smells	(6)
	Food types	(62)
	General terms	(64)
Total		**1,654**

2.2 Composing Definitions

The connoisseurs composed *le mot juste* following by the definition writing's principle [9], which containing no words more difficult to understand than the word defined.

Additionally, all definition of Thai dish names contains its ingredients for the benefits of computerized knowledge representation e.g., ontology or machine learning on the allergic purpose or diet. For example, people who is allergic to peanuts or avoids the high cholesterol of coconut milk as shown in Table 2.

Table 2. Composing definition

Language	Wordlist	Definition
[TH]	แกงมัสมั่นไก่	[เนื้อไก่ปรุงกับน้ำพริกแกงมัสมั่นใส่กะทิ มันฝรั่ง หัวหอมใหญ่ และถั่วลิสง]
	/kɛːŋ mássamàːn kài/	/núː :a kài pʰruŋ kàp náːm pʰrík kɛːŋ mássamàːn sài kàʔ tʰí manfáʔràːŋ hǔ :ahɔ̌ :myài lɛ́ʔ tʰùːalíʔsǒ ŋ/
[EN]	Chicken Massaman Curry	[Chicken in massaman curry cooked in *coconut milk* with potato, onion and *peanuts*.]

2.3 Distributing and Translating

Generally, translation is regarded as an act of communication in which the translator must overcome not only linguistic but also cultural barriers [10]. In this stage, ThaiEnglish wordlists were distributed to the translators who are the professors and native speakers of the CCTI, Faculty of Arts, Chulalongkorn University. Afterwards, wordlists and definitions were carefully translated into English, French, Italian, Spanish, German, Japanese and Chinese. Punctiliously, the translation process in this project follows the language manipulation's rule [11, 12] which it should not be to describe languages, but to analyze the ways in which thoughts are articulated in a message and can be reformulated in another language. Table 3 shows the standardized equivalent translation, operating on the discourse, of 'มัสมั่น' /mássamàːn/ which means the most famous Thai curry derived from Persian since the reign of King Narai the Great who was enthroned the Ayutthaya Kingdom from 1656 to 1688.

Table 3. Translation in eight languages

Language	Wordlist	Definition
[TH]	แกงมัสมั่นไก่	[เนื้อไก่ปรุงกับน้ำพริกแกงมัสมั่นใส่กะทิ มันฝรั่ง หัวหอมใหญ่ และถั่วลิสง]
[EN]	Chicken Massaman Curry	[Chicken in massaman curry cooked in coconut milk with potato, onion and peanuts.]
[FR]	*Poulet au Curry Massaman*	*[Poulet au curry massaman et lait de coco avec pommes de terre, oignon et cacahuètes.]*
[IT]	*Curry Musulmano di pollo*	*[Pollo in curry musulmano preparato in latte di cocco con patata, cipolla e arachidi.]*
[ES]	*Curry Massaman de pollo*	*[Pollo en curry massaman cocinado en leche de coco con patata, cebolla y cacahuetes.]*
[DE]	*Massaman-Curry mit Hähnchenfleisch*	*[Hähnchenfleisch in Massaman-Currypaste, gegart in Kokosmilch mit Kartoffeln, Zwiebeln und Erdnüssen]*
[JP]	鶏肉入りマッサマンカレー	[マッサマンペーストを使い、ジャガイモやピーナッツが入ったココナッツミルク入り甘口カレー]
[CH]	麻斯曼咖喱鸡肉	[鸡肉 煮熟麻斯曼咖喱 和椰奶加土豆洋葱花生]

2.4 Collecting and Verifying Data

After manually translated, all entries and definitions were collected and reorganized, were compared and verified. To make it suitable for the further research on Natural Language Processing, redacting and linking the data should be done. This process will be explained in Sect. 3, methodology.

2.5 Recording Native Speakers' Voice and Gathering Illustrations

Prior to recording the native speakers' voice, the wordlist was cleaned and reorganized alphabetically following by each languages' writing system, then the codes for voice recording were assigned as shown in Table 4. For example, 'EN1104007' the letter 'EN' stands for 'English', '11' means 'Category number 11' which is 'Dish name', while '04' refers to 'Subcategory number 14' which is 'Curry, Panaeng Curry', and finally '007' relates to the seventh entry, 'Chicken Massaman Curry'.

Table 4. The assigned codes for recording the native speakers' voice

Order	Subcategory	Word	Code
1	Curry, Panaeng Curry		EN1104001
2		Beef Massaman Curry	EN1104002
3		Cha-om Sour Curry with Fresh Prawn	EN1104003
4		Chicken Green Curry	EN1104004
5		Chicken Yellow Curry	EN1104005
6		Chicken in Jungle Curry	EN1104006
7		Chicken Massaman Curry	EN1104007
8		Chicken Red Curry with Fresh Bamboo	EN1104008
8		Chicken Spicy Curry with Wild Betel and Shoots	EN1104009
10		Chicken Spicy Curry with Preserved Bamboo Shoots	EN1104010

Following the codes above, then the native speakers' voices were recorded. Unlike voice recording process; however, the illustrations were collected as Thai words since Thai language is the pivot of all languages in this work. For example, a picture 'แกงมัสมั่นไก่.PNG' will link to the voice-recorded file 'EN1104007' and to the texts 'แกงมัสมั่นไก่', 'Chicken Massaman Curry', *'Curry Massaman de Poulet'*, *'Curry Musulmano di pollo'*,

'Curry Massaman de pollo', *'Massaman-Curry mit Hähnchenfleisch'*,' 鶏肉入りマッサマンカレー' and '麻斯曼咖喱鸡肉'.

2.6 Caveats

Systematic Transliteration: Some Thai words have to be transliterated into foreign languages, but unsystematic due to the way of each languages' pronunciation. For instance, 'มัสมั่น' /mássamà:n/ which means Massaman-curry was firstly transliterated *'Massamane'* in French to avoid the vowel *'ã'* of the letters 'an', the editors decided to vacate 'e' because its well-known finally.

Equivalence of Translation: some crash of translation should be verified. For example, Thai fine-grained words 'ทา' /tʰā:/, 'ชุบ' /tɕʰŭb/ and 'พอก' / pʰɔ̄ :k/ which only mean 'to coat' in English but inadequate information and Thai. They should be extended to 'to coat by applying oil or egg yolk on the surface' for 'ทา' /tʰā:/, while 'ชุบ' /tɕʰŭb/ should be 'to coat by dipping', and 'to coat with a thick layer of batter' for 'พอก' /pʰɔ̄ :k/. On the contrary, four English words - 'creamy', 'nutty', 'oily' and 'greasy' - can be translated to only one word in Thai, 'มัน' /mān/. It, in the same way, should be extended to 'มันแบบกะทิ' /mān bὲp kàʔ tʰí/, 'มันแบบถั่ว' /mān bὲp tʰù:a/, 'มันแบบน้ำมัน' /mān bὲp ná:m mān/ and 'มันแบบเลี่ยน' /mān bὲp lî:an/ respectively.

Syntactic Coherence: Especially for the Romance languages, French, Spanish and Italian, the syntactic coherence of translation should be concerned. For example, a French phrase *'Poulet au Curry Massamane'* and a Spanish phrase *'Curry de pollo Massaman'* which mean 'chicken Massaman curry' should be rephrased to *'Curry Massaman de Poulet'* and *'Curry Massaman de pollo'* to conform with an Italian phrase *'Curry Musulmano di pollo'*.

3   Methodology

Given the aforementioned building-process of multimodal language resource, all data were collected in seven tables for seven languages: English, French, Italian, Spanish, German, Japanese and Chinese, as Thai language being pivot and linkage. Minimal inconsistency and human-error found, the data should be adjusted. The enhanced architecture has twofold: data preparation and generating structural data as follows.

3.1   Data Preparation

The multi-language translators of CCTI working independently, little unintentional typos exist in four cases: (1) inconsistent spacebar and enter key, (2) spacebar insertion between comma, (3) non-equivalent strings, and (4) re-formatting. As Thai language is the pivot and linkage of all foreign languages in this project. Prior to mapping together, all data should be ameliorated as follows.

Inconsistent spacebar and enter key :there were multiple spacebar keys, juxtaposing with an enter key at the end of the source strings (Table 5). Also, in many cases, an enter key were also found at the end of target strings. This case, single spacebar key and no enter key must be kept.

Table 5. Inconsistent spacebar and enter key

source	target
ผักแขนง, _แขนงกลํ่าปลี_ _ _ _ ↵	ผักแขนง, _แขนงกะหลํ่าปลี↵
ˈpʰàk kʰàʔ něǩ ːŋ/,[space]/ kʰàʔ něǩ ːŋ kàʔ làm pliː/	ˈpʰàk kʰàʔ něǩ ːŋ/,[space]/ kʰàʔ něǩ ːŋ kàʔ làm pliː/[enter]ˈ
[space][space][space][space][enter]ˈ	
ˈcabbage shootsˈ	ˈcabbage shootsˈ

Spacebar Insertion Between Comma: In this work, commas (,) were used for separating the synonymous words. In many cases, no space was inserted next to the comma. To map source and target data, a spacebar, without enter key, must be added after the comma as shown in Table 6.

Table 6. Spacebar insertion between comma

source	target
เครื่องแกง,พริกแกง,น้ำพริกแกง⏎	เครื่องแกง,_พริกแกง,_น้ำพริกแกง
ꞏ/kʰrɯ̂ ːaŋ kɛːŋ/,/pʰrík kɛːŋ/,/náːm	ꞏ/kʰrɯ̂ːaŋkɛːŋ/,[space]/pʰrík
pʰrík kɛːŋ/[enter]ꞏ	kɛːŋ/,[space]/náːm pʰrík kɛːŋ/ꞏ
ꞏcurry pasteꞏ	ꞏcurry pasteꞏ

Table 7. Non-equivalent strings

source	target
-ข้าวน้ำพริกกะปิปลาทูทอด	ข้าวน้ำพริกปลาทูทอด
/kʰâːo náːm pʰrík kàʔ pìʔ plaː tʰuː tʰɔ̂ ːt/	/ kʰâːo náːm pʰrík plaː tʰuː tʰɔ̂ ːt/
ꞏrice with fried mackerel and <u>shrimp paste</u>	ꞏrice with fried mackerel and
chili dipꞏ	chili dipꞏ

Table 8. Re-formatting

Former format	New format	
Head word + Definition	**Head word**	**Definition**
		（鶏肉とにんにく、唐辛子、ホーリーバジルを炒めて目玉焼きを添えたごはん）
鶏肉のホーリーバジル炒め目玉焼き添えとご飯（鶏肉とにんにく、唐辛子、ホーリーバジルを炒めて目玉焼きを添えたごはん）	鶏肉のホーリーバジル炒め目玉焼き添えとご飯	
/toriniku no hōrībajiru itame medamayaki soe to gohan (toriniku to nin'niku, tōgarashi, hōrībajiru o itamete medamayaki o soeta gohan)/	/toriniku no hōrībajiru itame medamayaki soe to gohan/	/toriniku to nin'niku, tōgarashi, hōrībajiru o itamete medamayaki o soeta gohan/
		ꞏrice topped with minced chicken sautéed with holy basil and chilies served with fried egg, sunnyside up as a side dipꞏ
ꞏrice with sautéed chicken, holy basil and fried egg (rice topped with minced chicken sautéed with holy basil and chilies served with fried egg, sunny-side up as a side dish)ꞏ	ꞏrice with sautéed chicken, holy basil and fried eggꞏ	

Non-equivalent Strings: Some entries share the same index, but unequal strings. In Table 7, a Thai word 'กะปิ' /kà? pì?/ 'shrimp paste' exists in the source file, but the target file does not. In this case, the non-equivalent strings should be verified by the experts which one being accurate, then collimated afterwards.

Re-formatting: The head word, in some files, was juxtaposed with its definition. The Reformatting must be done to divide word entry and definition shown in Table 8.

3.2   Generating Structural Data

According to the designed data taxonomy consisted of hierarchy – fourteen categories together with nineteen subcategories – and the lexical entries, we transformed the data structure into tree structure (Fig. 2), then assigned and linked the conceptual nodes in that tree structure, to encourage efficiency, before utilizing these benefits in the searching tool such as the mobile application.

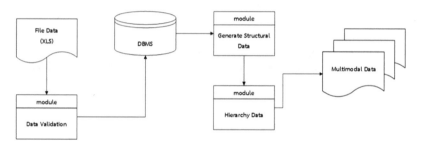

Fig. 2. Tree structure transformation

4   Scenario Settings and Discussion

Scenario 1. New Knowledge Discovered via Inheritance: Through the designed taxonomy, fourteen categories with hypernym-hyponym relation were created. An advantage of these semantic relation is to inherit whose semantic features. If the nutrition information or other features are added, we can create the lexical database which contains this information. For example, If we defined that 'แกงกะทิ' /kɛ:ŋ kà? tí?/ or the dishes contained coconut milk is a hypernym which has 'แกงมัสมั่น' /kɛ:ŋ mássamà:n/ 'Massaman curry', 'แกงเขียวหวาน' /kɛ:ŋ kʰĭao wă:n/ green curry', 'แกงพะแนง' /kɛ:ŋ pʰáne:ŋ/ 'Panaeng curry', 'แกงคั่ว' /kɛ:ŋ kʰû:a/ 'Gaeng Kua curry' and 'แกงกะหรี่' /kɛ:ŋ ka?rì:/ 'yellow Curry' as the hyponyms. When 'กะทิ' /kà? tí?/ 'coconut milk' is a property of 'แกงกะทิ' /kɛ:ŋ kà? tí?/,, every Thai-style curry contained coconut milk as an ingredient can be retrieved.

Scenario 2. Time Effective: Due to the combination of texts, images and speeches in the relational database, the information retrieval through any application should be designed to support all three kinds. To evaluate the efficiency and to envisage the searching tool, mobile application, in future, then, we developed and examined two information retrieval methods: (1) through relational database and (2) though noSQL. The result in Fig. 3 shows that the retrieval via noSQL is faster than the way of relational database. When 1,500 entries examined, the time effective of noSQL is only 0.031 s as the hierarchical structure has completely been made, while the relation database dissipates 0. 033 s since the designed index structure. Additionally, the size of noSQL is smaller, 1490 KB, while the bigger relation database, 16.4 MB. The former method is suitable for mobile application than the latter.

Scenario 3. Multimodal Learning: To construct the multimodal language resource for Thai cuisine, the texts, images and speeches were assembled and stored in the designed taxonomy. All data conjoined, we can apply this benefit for the further research, for example, using this united data on the machine learning for the artificial intelligence's information retrieval. When searching for a Thai word 'ถั่วลิสง' /tʰùːalíʔsǒ ŋ/ or 'peanuts' in English, even a German *'Erdnüssen'* or a French *'cacahuètes'*, It does not only retrieve the meaning and translation of; pictures and pronunciations of peanuts but the dishes containing peanuts. Similarly, the shape – form – color of an unknown food image or speech can be differentiated to investigate the dish name with nutrition, ingredients and cooking techniques.

5   Conclusion and Future Works

In this paper, we describe the current status of the first standardized multimodal language resource containing texts, images and speeches. Thoroughly translated from Thai to seven foreign languages: English; Italian; Spanish; and, Chinese; German; Japanese, plus French, 1,654 parallel lexical entries with definitions were accumulated. All data were connected and stored in the transformed tree structure with fourteen categories and nineteen subcategories.

In future, utilizing an advantage of the finding in this project is to develop a searching tool, a mobile application, manifesting lexical data, illustrations and emitting recorded human's voices. Availing the multimodal language resource, the application can represent and suggest the personally healthy dietary pattern of Thai food tailored to personal conditions and preferences via searching both texts, images and speeches. For the new items to be added in future, the first priority for categorizing the dish is the appearance or visual aspect of thing. For example, a fictitious thing 'coconut rice noodle' will be categorized as noodle.

Acknowledgements. The aim of this project is to celebrate H.R.H. Princess Maha Chakri Sirindhorn's the sixtieth-years-birthday anniversary. The financial support for this study was provided by Faculty of Arts, Chulalongkorn University. Additionally, the copyright of the Thai food terms belongs to Chalermprakiat Center of Translation and Interpretation, Faculty of Arts, Chulalongkorn University.

References

1. Thai Food to The World. http://www.thaifoodtoworld.com/
2. Thai Food Secrets. http://www.modernthaifood.com/pdfs/thai_food_secrets_index.pdf
3. Modern Thai Cooking. http://www.simplyasiacanada.ca/public/simplyasiaca/books/modernthaicooking.pdf
4. ABC
5. Smola, A., Vishwanathan, S.V.N.: Introduction to Machine Learning. Oxford University Press, Oxford (2008)
6. Shalev-Shwartz, S., Ben-David, S.: Understanding Machine Learning: From Theory to Algorithms. Cambridge University Press, Cambridge (2014)
7. Murphy, K.P.: Machine Learning. The MIT Press, Cambridge (2013)
8. Domingos, P.: A few useful things to know about machine learning. Commun. ACM **55**(10), 78–87 (2012)
9. Atkins, B.T.S., Rundell, M.: The Oxford Guide to Practical Lexicography. Oxford University Press, New York (2008)
10. Šarčević, S.: Lexicography and translation across cultures. In: Translation and Lexicography: the Euralex Colloquium, 2–5 July 1987, pp. 211–221. Innsbruck (1987)
11. Delisle, J.: Translation: An Interpretive Approach. University of Ottawa Press, Ottawa (1988). Translated by Patricia Logan and Monica Creery
12. Levý, J.: The Art of Translation. John Benjamins, Amsterdam, Philadelphia (2011). Translated by Patrick Corness. Edited with a critical Foreword by Zuzana Jettmarová

Cloud-Based Services for Cooperative Robot Learning of 3D Object Detection and Recognition

Parkpoom Chaisiriprasert[1(✉)], Karn Yongsiriwit[1],
Apiporn Simapornchai[2], and Matthew Dailey[2]

[1] College of Information and Communication Technology, Rangsit University,
Lak Hok, Thailand
{parkpoom.c,karn.y}@rsu.ac.th
[2] Computer Science and Information Management,
Asian Institute of Technology, Khlong Nueng, Thailand
apiporn.sima@gmail.com, mdailey@ait.ac.th

Abstract. Some of the problems preventing the widespread adoption of advanced service robots are (1) cost, (2) power consumption, (3) perceptual capabilities, (4) knowledge management, (5) reasoning capabilities, and (6) compute power. Improvement along these dimensions will make adoption of service robots with advanced capabilities much more widespread. We propose the use of real time cloud computation as one means to enable these improvements. This paper presents a case study on the use of a cloud computing platform to support robotics applications, allowing robots to offload heavy compute tasks such as machine vision to cloud infrastructure. We specifically aim to give a widely distributed group of robots the ability to learn 3D objects cooperatively for detection and recognition. Participating robots can share their knowledge with others via our cloud-based services. Experiments with a proof of concept prototype demonstrate the feasibility of the use of cloud platforms to deliver improved perceptual, knowledge management, and reasoning capabilities while keeping the cost and power consumption of the robot low.

Keywords: Cloud-based services · Machine vision
3D object detection and recognition · Cooperative robot learning

1 Introduction

Compute-intensive perceptual reasoning tasks such as object detection and recognition are basic behaviors that service robots [1] must perform in order to assist humans over a broad variety of different day-to-day activities. Effective object detection and recognition requires a great deal of data storage and compute power, more than would be typical of an embedded robot control system. If it were possible for the embedded system to offload the necessary storage and compute capabilities to a more cost effective centralized infrastructure, it may be possible to introduce service robots with vastly improved perceptual reasoning capabilities at very low costs. This idea has led us to consider the applicability of cloud computing [2] to support service robots,

© Springer Nature Switzerland AG 2019
T. Theeramunkong et al. (Eds.): iSAI-NLP 2017, AISC 807, pp. 14–24, 2019.
https://doi.org/10.1007/978-3-319-94703-7_2

especially in perceptual reasoning tasks such as object detection and recognition. There has been some research on solutions to open problems in the use of multiple networked robots, such as localization and mapping, cooperative robot learning, skills for service robots, and knowledge sharing.

However, there is still a large gap between the abilities of individual or small groups of networked robots and what is required for truly useful multi-function autonomous service robots. Connecting multiple robots with a network helps us to share and pool resources, but architectures for Internet-scale distributed robot learning have not been extensively explored.

In this paper, we aim to extend the capabilities of networked service robot systems, while reducing the cost, on-board computational power, and storage required for the individual service robots, with proof of concept prototype of cloud-based services for the specific case study of 3D object detection and recognition. Experiments with the prototype demonstrate the feasibility of the use of cloud platforms to deliver improved perceptual, knowledge management, and reasoning capabilities of individual robots while keeping the cost and power consumption of the robot low.

The rest of this paper is organized as follows. Section 2 presents the related work. Section 3 describes an architecture for cooperative robot learning through cloud computing, with a specific case study of object detection and recognition for service robots. Section 4 details our approach for object training, detection and recognition. Use cases and experiments are given in Sect. 5. Finally, Sect. 6 concludes the paper with an outlook to future work.

2 Related Work

Several research groups have introduced of cloud computing into robotic systems. Jordan et al. [3] and Kehoe et al. [4] survey the rising prospects of cloud robotic applications and for a comprehensive survey. The systems developed to date can generally be divided according to the traditional cloud computing service models "platform as a service" (Paas) and "software as a service" (Saas), or into a new "robot as a service" (RaaS) category.

Yinong et al. [5] were the first to propose the RaaS model. They provide a common service standard, development platform, and execution infrastructure. Zhihui et al. [6] also present a RaaS system, a cloud center providing a utility service through which users can access multi-robot resources at minimal cost using a robot scheduling algorithm.

At the PaaS level, Agostinho et al. [7] present the REALabs cloud computing platform for supporting network robotics applications. The system includes a VM management service interface for Virtual Box. It allows users to manage of their own VMs. The platform provides a set of services for robotics applications such as access control, federated authentication, and resource protection. This work can be used by other robotic applications to solve problems such as knowledge sharing.

At the SaaS application level, Arumugam et al. [8] propose a cloud framework for service robots called DAvinCi. They deploy a parallel implementation of FastSLAM on Hadoop and the Map-Reduce model. This work demonstrates the ability to efficiently build a map of a large area by sharing data across the robots. Another SaaS example is the senior companion robot system (SCRS) [9]. The authors describe a method for human-robot interaction based on cloud computing that offers a service for remote robot control. SCRS can support and assist seniors in daily activities using speech for home appliance control, reading books and so on. Tenorth et al. [10] present a knowledge-enabled cloud robotics application using a ubiquitous network robot platform. Their system is designed for the exchange of knowledge between robots. They implemented knowledge representations using the Web ontology language in the ROBOEARTH system. The case study scenario is a recommendation robot in a convenience store. The robot's sensors include a laser scanner for tracking customers and an RFID tag reader for detecting the objects that have been picked up.

Moving toward PaaS and SaaS for perceptual reasoning, especially machine vision, according to Kehoe et al. [11], Willow Garage's PR2 performed a complex and difficult task with cloud based support: the robot acquires 2D and 3D images of an object, sends the 2D image for processing by an object recognition server. If the server based application can identify an object in the image, it returns a 3D CAD model to the robot. The robot then uses the measured 3D point cloud data with the 3D CAD model to determine appropriate grasping strategies and stores the results in the cloud for future use.

Turnbull and Samanta [12] propose a small scale cloud robotics application for formation control that offloads computationally intense perceptual tasks to the cloud infrastructure. Their system consists of multiple vision acquisition sensors providing image data to the cloud system. The cloud system estimates the location and behavior of the robots in the formation using recognition algorithms, then it chooses appropriate commands to send back to the robots.

Our approach to offloading perceptual reasoning capabilities to the cloud is based on a combination of the PaaS and SaaS models. In the next section, we provide details of our system architecture. The platform extends the Robot Operating System (ROS) framework to a cloud computing environment, with the addition of on-demand allocation of cloud resources to specific users' robotics applications. As a case study, we present an implementation of 3D object detection and recognition utilizing the cloud platform.

3   System Architecture

In this section, we propose an architecture for large-scale distributed robot learning through cloud computing, with a specific case study of object detection and recognition for service robots.

Motivation Example

We propose the use of a cloud platform not only to share and pool resources, but to also provide cloud-based services that offer the means to cooperatively solve perceptual reasoning tasks among multiple robots. We illustrate our proposal with a motivation example in Fig. 1. Let us consider two known objects (A and B) are shot using different robots with cameras. The structure and appearance of each object is then captured and represented by *Point Cloud Data* (PCD) [15] (i.e., a set of point specifying position in space). Such data is then uploaded to the cloud platform through *Object Training Service*, hence constructing a shared training set for object detection and recognition, namely *Object Templates*. We take into account multiple object templates for a single object to represent its different viewpoints which is necessary for identifying asymmetric objects (e.g., A). On the other hand, another robot wants to identify an object A in the target scene X represented by PCD. It can be done by using *Object Detection and Recognition Service* to perform matching of the target scene with *Object Templates* specifically for the object A. Multiple object template matching processes run in parallel in the cloud platform to reduce the time required to find the best matching. Finally, the best matching template returns to the robot for identifying the position and orientation of the object A in the scene.

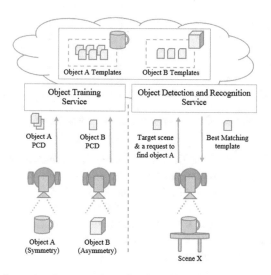

Fig. 1. Scenario of cooperative robot learning using the cloud platform

Architecture

Figure 2 gives a high level overview of the platform. The architecture assumes a private cloud. Although a very similar system could be constructed using a public cloud such as Amazon EC2 controlled through an interface such as Amazon Web Services, with a private cloud we have full control over the hardware, virtualization, and software stacks. This allows us to offer on-demand computing resources to robotics applications, by provisioning and managing a large network of specialized virtual machines. The architecture of the cloud platform composes of four main components as follows:

Extended ROS Framework. We extend the ROS framework to a cloud computing environment. Thus, robots can access the cloud-based services over the internet by using ROS [14]. Concretely, we create 3 cloud-based services: (*i*) *Robot Registration Service* handles initial registration of robots to the cloud platform. (*ii*) *Object Training*

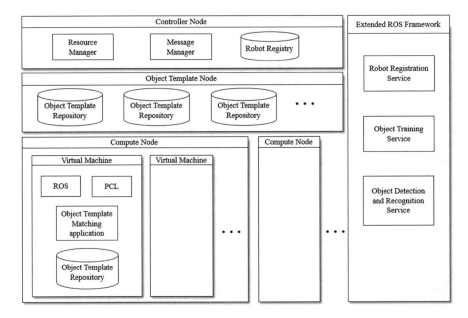

Fig. 2. Cloud platform based on ROS framework

Service receives the Point Cloud Data (PCD) representing the structure and appearance of a training object acquired from networked robots, hence constructing a training set for a specific object, so called object templates. (*iii*) *Object Detection and Recognition Service* receives a target scene and a request to identify an object in the scene. Such scene is also represented as PCD. The service therefore performs matching of the target scene with previously captured object templates of the requested object, thus responses to the robot with the best matching template.

Controller Node. The *Controller Node* distributes computational tasks to multiple *Compute Nodes* running object template matching applications. It consists of 3 sub-components: (*i*) *Robot Registry* stores robot identification for accessing the cloud platform. (*ii*) *Message Manager* handles communication between components in the cloud platform using ROS proxy and messages. (*iii*) *Resource Manager* dynamically creates multiple *Compute Nodes* with virtual machine instances. Thereafter, it distributes matching tasks to multiple virtual machines for running in parallel which help reducing the time required to find the best matching template.

Object Template Node. We deploy a highly scalable repositories which provides support for storing and retrieving object templates on the cloud platform. Each *Object Template repository* stores multiple PCD of a specific object. We consider multiple PCD for a single object to represent its different object viewpoints. This helps improving object template matching results, especially for asymmetric objects.

Compute Node. *Compute Nodes* support management and automation of pools of computer resources via virtualization technologies. We have developed virtual machine instances responsible for performing cooperative learning for object detection and recognition. Concretely, *Compute nodes* are developed based on ROS and are dynamically created by the *Controller Node* when performing object template matching. Each node contains an *Object Template Matching application* and a local *Object Template Repository* which is partially retrieved from *Object Template Node* to offloads the computationally intense of PCD matching. Such matching is likewise based on Point Cloud Library (*PCL*) [15] functions.

4 Cooperative Robot Learning Approach

Here we provide a description of the cooperative object learning, detection and recognition approach built on top of the previously-described cloud-based platform. Figure 3 shows a sequence diagram of the cooperative learning approach. We assume for the time being that each robot participating in the distributed system has a camera and a Wi-Fi connection to the Internet. The method is divided into two phases: (*i*) object training, and (*ii*) object detection and recognition.

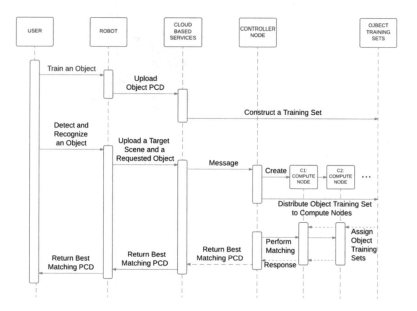

Fig. 3. Sequence diagram for object detection and recognition applications

Object Training

We propose to collect multiple PCD for an object, and thus constructing a training set for object detection and recognition, namely *object templates*. Such PCD are uploaded and stored in an object template repository in the cloud platform through the *Object Training Service*. We capture multiple object templates of an object to represent different viewpoints of an object as shown in Fig. 4. Thus, this helps improving object template matching results for identifying asymmetric objects.

Fig. 4. Multiple object templates representing different viewpoints of an asymmetric object

A user can create and upload PCD of an object to the cloud platform via the *Object Training Service* using a simple application implemented as a ROS package. Refer to Fig. 5. Such application allows users to detect and subtract any large planar surface in order to segment objects from the background (usually a table or other at surface).

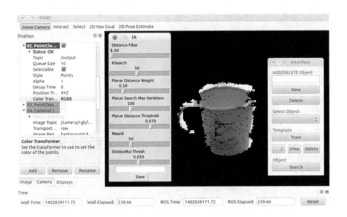

Fig. 5. An application for capturing and uploading PCD

Object Detection and Recognition

To detect and recognize an object using a robot with a camera, we firstly acquire an image to get a Point Cloud Data (PCD) (Fig. 6(a)). We remove outlier points using statistical outlier removal, thus we segment objects from the background (Fig. 6(b) and (c)). Finally, we use the *Object Detection and Recognition Service* of the *extended ROS framework* to identify an object using template matching using Sample Consensus Intial Alignment (SAC-IA) algorithm [16]. Given a set of previously-captured template of an object, we can thus determine its position and orientation in the scene. Concretely, we take a depth image containing an object and try to fit some previously captured template to the object. This works well for getting the position and orientation of the object in a cluttered scene as shown in Fig. 7. However, finding the best match in the point cloud representation is computationally expensive, because the number of object templates required is relatively large. We therefore propose the use of cloud-based parallel computing to reduce the time required to compute the best match.

(a) Acquire an image   (b) Subtract background   (c) Capture the PCD

Fig. 6. Capturing PCD from an acquired image

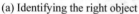

(a) Identifying the right object   (b) Identifying the left object

Fig. 7. Examples of object identification in a target scene

In case the robot cannot respond to request of user, for example when the object is unknown, then the robot should interact with the user and the learning module, uploading the acquired PCD to the cloud. In addition, a robot may learn to improve the training sets in the cloud platform and try to respond to the request of the user at the same time. Cooperative learning between robots is a simple matter of sharing templates for specific objects.

5   Experiments

We configured an experimental hardware setup modelling a complete cloud installation according to the topology shown in Fig. 8. In this topology, the controller node, the network node, and the compute nodes are on separate physical cores. We use one cloud controller node that can distribute computational tasks to the four compute nodes running the object template matching application, and one network node to manage message sending, receiving, and forwarding. We first created initial sample object template repositories for the experiments containing 3D images in PCL format of a set of household objects. We constructed a dataset consisting of 10 templates for 100 different objects. Figure 9 shows an example of the objects.

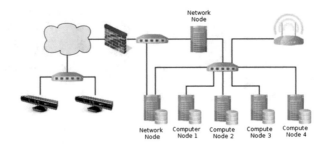

Fig. 8. Hardware setup for experiments.

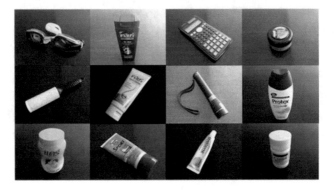

Fig. 9. Example household objects used in experiments.

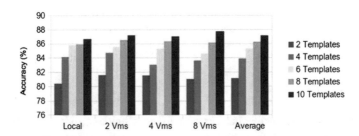

Fig. 10. Object recognition accuracy in Experiment I.

We then performed two experiments to evaluate the accuracy and the performance of the identification of objects in the data set based on novel views not present in the original repository. In Experiment I, we evaluated the accuracy of our approach for different number of object templates and number of virtual machines. As shown in Fig. 10, one would expect accuracy increases with the number of object templates in the repository but is unaffected by the number of virtual machines used. In Experiment II, we evaluated the latency for different number of object templates and number of virtual machines. As shown in Fig. 11, increasing the number of object templates also increases latency but this effect can be mitigated by adding additional virtual machines. In practice, an effective balance between accuracy, latency, and resource utilization should be found.

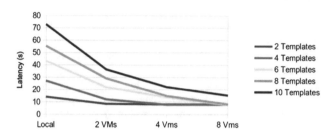

Fig. 11. Object recognition latency in Experiment II.

6   Conclusion

In this paper, we have presented a combination of the PaaS and SaaS service models for cloud computing applied to recognition of 3D objects by a possibly large and distributed group of service robots. Experiments demonstrate the feasibility of the concept. Cloud-based parallel computation dramatically reduces the time required to compute the best match between a repository of object templates and a given scene, and the position and orientation of the object in the scene can be retrieved. The main practical limitations of the approach are that it does not perform well in identifying specific objects with similar shapes, and recognition latency needs to be improved. We plan to improve the result by using 3D key points and improved appearance modelling for identification of specific objects.

References

1. Cao, Y.U., Fukunaga, A.S., Kahng, A.B., Meng, F.: Cooperative mobile robotics: antecedents and directions. In: IEEE/RSJ International Conference on Intelligent Robots and Systems 1995, Human Robot Interaction and Cooperative Robots, Proceedings 1995, vol. 1, pp. 226–234, August 1995
2. Armbrust, M., Fox, A., Griffith, R., Joseph, A.D., Katz, R.H., Konwinski, A., Lee, G., Patterson, D.A., Rabkin, A., Stoica, I., Zaharia, M.: Above the clouds: a berkeley view of cloud computing. Technical Report UCB/EECS-2009-28, EECS Department, University of California, Berkeley, February 2009
3. Jordan, S., Haidegger, T., Kovacs, L., Felde, I., Rudas, I.: The rising prospects of cloud robotic applications. In: 2013 IEEE 9th International Conference on Computational Cybernetics (ICCC), pp. 327–332, July 2013

4. Kehoe, B., Patil, S., Abbeel, P., Goldberg, K.: A survey of research on cloud robotics and automation. IEEE Trans. Autom. Sci. Eng. **12**(2), 398–409 (2015)
5. Chen, Y., Du, Z., García-Acosta, M.: Robot as a service in cloud computing. In: Fifth IEEE International Symposium on Service Oriented System Engineering (SOSE) 2010, pp. 151–158, June 2010
6. Du, Z., Yang, W., Chen, Y., Sun, X., Wang, X., Xu, C.: Design of a robot cloud center. In: 10th International Symposium on Autonomous Decentralized Systems (ISADS) 2011, pp. 269–275, March 2011
7. Agostinho, L., Olivi, L., Feliciano, G., Paolieri, F., Rodrigues, D., Cardozo, E., Guimaraes, E.: A cloud computing environment for supporting networked robotics applications. In: IEEE Ninth International Conference on Dependable, Autonomic and Secure Computing (DASC) 2011, pp. 1110–1116 (2011)
8. Arumugam, R., Enti, V.R., Bingbing, L., Xiaojun, W., Baskaran, K., Kong, F.F., Kumar, A.S., Meng, K.D., Kit, G.W.: DAvinCi: a cloud computing framework for service robots. In: IEEE International Conference on Robotics and Automation (ICRA) 2010, pp. 3084–3089, May 2010
9. Chen, Y.-Y., Wang, J.-F., Lin, P.-C., Shih, P.-Y., Tsai, H.-C., Kwan, D.-Y.: Human-robot interaction based on cloud computing infrastructure for senior companion. In: TENCON 2011-2011 IEEE Region 10 Conference, pp. 1431–1434, November 2011
10. Tenorth, M., Kamei, K., Satake, S., Miyashita, T., Hagita, N.: Building knowledge-enabled cloud robotics applications using the ubiquitous network robot platform. In: IEEE/RSJ International Conference on Intelligent Robots and Systems (IROS), pp. 5716–5721, November 2013
11. Kehoe, B., Matsukawa, A., Candido, S., Kuffner, J., Goldberg, K.: Cloudbased robot grasping with the Google object recognition engine. In: IEEE International Conference on Robotics and Automation (ICRA) 2013, pp. 4263–4270, May 2013
12. Turnbull, L., Samanta, B.: Cloud robotics: formation control of a multi robot system utilizing cloud infrastructure. In: Southeastcon 2013 Proceedings of IEEE, pp. 1–4, April 2013
13. Sefraoui, O., Aissaoui, M., Eleuldj, M.: Openstack: toward an open-source solution for cloud computing. Int. J. Comput. Appl. **55**(3), 38–42 (2012)
14. Quigley, M., Conley, K., Gerkey, B., Faust, J., Foote, T.B., Leibs, J., Wheeler R., Ng, A.Y.: ROS: an open-source robot operating system. In: ICRA Workshop on Open Source Software (2009)
15. Rusu, R.B., Cousins, S.: 3D is here: Point cloud library (PCL). In: IEEE International Conference on Robotics and Automation (ICRA), Shanghai, China, May 2011
16. Open Perspective Foundation. Aligning object templates to a point cloud. http://pointclouds.org/documentation/tutorials/template_alignment.php. Accessed 01 Apr 2017

Detection of Normal and Abnormal ECG Signal Using ANN

Sourav Mondal$^{(\boxtimes)}$ and Prakash Choudhary

Department of Computer Science and Engineering, NIT Manipur,
Manipur, Imphal, India
{mondal.sourav2010, choudharyprakash87}@gmail.com

Abstract. The normality & abnormality of the heart is normally monitored by ECG. Several algorithms are proposed to classify ECG signals. In this paper, discrete wavelet transform is used for extracting some statistical features and Multilayer perceptron (MLP) neural network with Back-propagation performs the classification task. Two types of ECG signals (normal and abnormal) can be detected in this work from each database. The records from MIT-BIH Arrhythmias and Apnea ECG database from physionet have been used for training and testing our neural network based classifier. 90% healthy and 100% abnormal are detected in MIT-BIH Arrhythmias database with the overall accuracy of 94.44%. In Apnea-ECG database, 96% normal and 95.6% abnormal ECG signals are detected and achieves 95.7% classification rate.

Keywords: ECG · Daubechies · Discrete wavelet transform · Neural network
MIT-BIH Arrhythmia database · Apnea ECG database

1 Introduction

The Electrocardiogram (ECG) signals are the diagnostic tool that graphically measures the electrical activity of the heart over time in detail. Different persons have different cardiac features which are unique in nature. To detect the regularity and irregularity of the heartbeat Electrocardiogram is one of the classical diagnosis [1]. This is a unique tool for diagnosing several cardiac diseases like Tachycardia, Cardiac arrhythmias, and other diseases or abnormalities.

The ECG signal composed of many waves namely P, Q, R, S, and T. These are the essential characteristics of ECG. R peak or QRS complex is the highest peak of the ECG signal. The QRS complex represents the depolarization of the right and left ventricles of the human heart.

The shape of an ECG signifies vital hidden information in its structure. The amplitude and duration of each wave in ECG signals are often used for the manual analysis. Thus the volume of the data being enormous and the manual analysis is a very time consuming task. Naturally, the possibility of the analyst missing vital information is high. Therefore, the medical diagnostic can be performed using computer-based analysis and classification techniques [2].

© Springer Nature Switzerland AG 2019
T. Theeramunkong et al. (Eds.): iSAI-NLP 2017, AISC 807, pp. 25–37, 2019.
https://doi.org/10.1007/978-3-319-94703-7_3

ECG arrhythmia affects the heart because of it the heart beats either slow or fast, and the heart beat is also irregular. A doctor can detect an irregular through an electrocardiogram (ECG). There are many symptoms of an arrhythmia which include.

Palpitations (a feeling of skipped heart beats, fluttering or "flip-flops," or feeling that our heart is "running away"), Pounding in our chest, Dizziness or feeling lightheaded, Fainting, Shortness of breath, Chest discomfort, Weakness or fatigue(feeling very tired) [3].

There are millions of people affected by sleep apnea worldwide. It is a common disorder which is very serious, about which a lot of people are not aware. According to the National Sleep Foundation, sleep apnea affects more than 18 million Americans. Sleep apnea is seen more frequently among men than women. Sleep apnea causes the blockage of airway while sleeping which results in reduction of air in blood, which ultimately results in high blood pressure, heart disease and stroke all of these are life threatening diseases. Sleep disturbances and repeated reduction in blood oxygen levels result in excessive daytime sleepiness, reduced the quality of life and impaired cognitive function such as memory loss and poor concentration [4].

Apnea is adapted from the Greek word which means without breath and sleep apnea refers to pauses in breathing that occurs during sleep. Loud snoring is one of the primary symptoms of sleep apnea disease. There are many sleep apnea symptoms, common symptoms include waking up with a very sore or dry throat, occasionally waking up with a choking or gasping sensation, sleepiness or lack of energy during the day, sleepiness while driving, morning headaches, restless sleep, forgetfulness, mood changes, decreased interest in sex and Recurrent awakenings or insomnia [4].

There are two types of sleep apnea:

(a) Obstructive sleep apnea **(OSA):** The more common of the two forms of apnea, it is caused by a blockage of the airway, usually when the soft tissue in the back of the throat collapses during sleep.

(b) Central sleep apnea: Unlike OSA, the airways is not blocked, but the brain fails to signal the muscles, due to instability in the respiratory control center [5].

The different types of ECG signal taken from physionet database from different subjects used in this work are shown in Figs. 1 and 2.

Rest of the paper is organized in 6 sections. Section 2 describes the methodology; it has two Subsects. 2.1 and 2.2 which describe the data collection and preprocessing steps. Section 3 describes the features extraction stage with 3.1 and 3.2 subsections where we explain the process of discrete wavelet transform and statistical feature extraction method. Principal component analysis method describe in Sect. 4. Section 5 describes the classification method followed by results and conclusion in Sects. 6 and 7.

2 Methodology

The method used in this work involves Data Collection, Pre-processing, Feature extraction and the Classification stages. Each stage will be explained in the next subsections (Fig. 3).

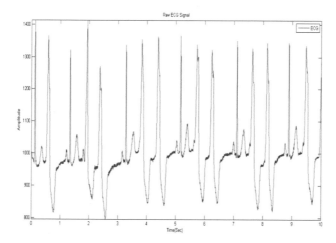

Fig. 1. 10 s raw ECG signal of Arrhythmias patient.

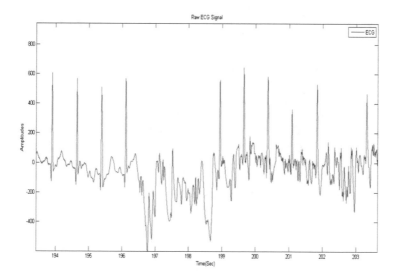

Fig. 2. Raw ECG signal of Sleep-Apnea patient (few samples from 1st 1 h recording).

2.1   Data Collection

In this work, we used ECG signals taken from public online database called PhysioNet. In this study the source of the ECG data used for training and testing is MIT-BIH Arrhythmia and Apnea-ECG database. The MIT-BIH Arrhythmia database contains 48 recordings of both routinely and some complex arrhythmias sampled at 356 Hz of 30 min duration from 24 h recording with two channels obtained from 47 patients. Only 1 channel of 10 s for each record is used in this work. The MIT-BIH Arrhythmia database is divided into two classes normal and abnormal. We used this database for normal and abnormal detection.

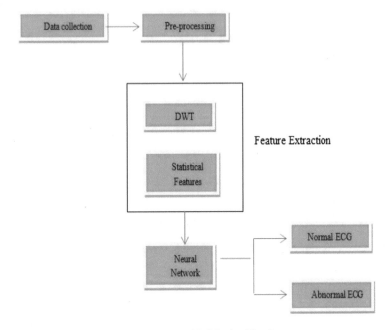

Fig. 3. Block diagram of ECG classification system.

The Apnea-ECG database contains 70 records, separated into learning set of 35 records. Recordings vary in length from slightly less than 7 h to nearly 10 h each. The sampling frequency used for ECG acquisition was 100 Hz, with 16-bit resolution. The subjects of these recordings were men and women between 27 to 63 years of age [6]. Total 1 h recording used for each record in this work. This database is divided into two classes Normal and abnormal. Using Apnea-Hypopnea index (AHI), we separate the two classes. If AHI >= 10, the patient has placed in apnea class or abnormal class and AHI < 10 the ECG record come under normal class.

2.2 Preprocessing

This stage is very important to obtain useful information from a raw signal. The purpose of Pre-processing stage is the removal of noise such as baseline drift, powerline noise, electrode contact noise, instrumentation noise generated by electronic devices. In this stage, the signals are filtered by using the band-pass filter, Butterworth Filter, and Chebyshev Type II filter. This stage improves the classification accuracy. The design of band-pass filter with Butterworth's uses the following parameter for the analysis of normal ECG signal from MIT-BIH Arrhythmias database in Matlab:

$$[n, Wn\ ] = buttord\ (Wp, Ws, Rp, Rs)$$

The above function returns the lowest order, n, of the digital Butterworth filter with no more than Rp dB of passband ripple and atleast Rs dB of attenuation in the stopband. The cut-off frequency Wn is also returned. The cut-off frequency must be between 0 to 1. After then, we used the output arguments in butter() function to obtain the Butterworth filter transfer functions coefficients. Then it is converted to secondorder-section for stability using tf2sos() function. This function founds an L-by-6 matrix which contains the coefficients of second-order-section for filtering with ECG_data to obtain the filtered ECG_data. That filter has removed the baseline drift at the low end and noise at the high end.

Similarly, Chebyshev Type II bandpass filter and 60 Hz noise filter (notch filter) was used for baseline removal and powerline noise for ECG with arrhythmia (abnormal). At first we removed the baseline drift using Chebyshev Type II bandpass filter and then we used Chebyshev Type II 60 Hz notch filter to remove the higher frequencies.

Chebysehev type II high pass filter was used for the removal of baseline noise from Apnea-ECG database with 0.8 cut-off frequency. The below figures are the raw and filtered some ECG samples of different subjects (Fig. 4).

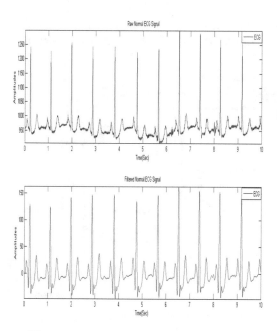

Fig. 4. Raw and filtered normal ECG signal.

3 Feature Extraction

In the feature extraction stage DWT features are determined. In this stage ECG features are extracted using selecting 10 s of ECG signal from each record of MITBIH Arrhythmias database and 1 h ECG record from Apnea ECG database.

3.1 Discrete Wavelet Transform

The wavelet transform can be applied in both continuous-time signal and discretetime signal. Discrete Wavelet Transform consists of a series of filtering and subsampling. The signal x(n), where n is the number of samples, are passed through the LowPass Filters (LPF) and High- Pass Filters (HPF) and down-sampling the output signal by 2, so that each frequency band contains n/2 samples. The two types of filtering occur at the same time. The number of coefficients varies with the decomposition level and Daubechies wavelet used. At each level of decomposition two sets of coefficients are calculated detail coefficients and approx coefficients.

This decomposition is repeated in a recursive cascade structure of the binary tree with nodes known as the filter bank. Figure 7, graphically represents this process, where x(n) is the original signal, h(n) and g(n) are the high pass and low pass filter impulse response respectively. Reconstruction occurs with filtering and up-sampling of the resultant coefficients. DWT produces different wavelet families like Daubechies (db), Haar, coiflets, etc. Among each family of wavelets, there is wavelet subclass defined by the no. of coefficients and the level of iterations [7]. In this work, we select Daubechies wavelet due to its resemblance of an ECG signal (Fig. 5).

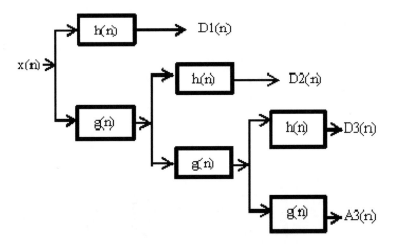

Fig. 5. Three level wavelet decomposition tree.

Various wavelet families are defined in the literature. Daubechies wavelets are the most popular wavelets are used in different applications. The wavelets filters are selected based on their ability to analyze the signal and their shape in an application [7]. Different features can be calculated using wavelet coefficients. We can apply the wavelet transform on the ECG signal and then it is converted to the number of wavelet coefficients or parameters, which characterizes the behavior of the ECG signal.

3.2 Statistical Features

In this stage, the ECG features are extracted using selecting 10 s of an ECG record from MIT-BIH arrhythmia database and 1 h record from Apnea-ECG. We have used DWT to extract some statistical parameters or features from DWT coefficients. As already mentioned above there are many wavelet filters to apply on a signal. The following statistical features were used:

- Standard deviation of the wavelet coefficients in each sub-band.
- Energy of the wavelet coefficients in each sub-band.
- Entropy of the wavelet coefficients in each sub-band.
- Mean of the wavelet coefficients in each sub-band.
- Median of the wavelet coefficients in each sub-band.

In this work, Daubechies wavelet of order 2 (db2) are used. The total 10 s 3600 samples of each ECG data record from MIT-BIH arrhythmia database. To reduce the volume of data, the sample was partitioned into 4 windows. Each window contains 900 samples. We performed the DWT in these subsamples and calculate the statistical features from these windows. The DWT was performed at 3 levels and resulted in five sub-bands: D1–D3 (detail coefficients) and A3 (approximation coefficients). For each of these sub-bands, we extracted 20 attributes per sample window. Total 20 * 4 = 80 attributes are obtained from 4 windows. Then the size of the feature vector was 45×80.

Similarly, the ECG signals of Apnea ECG database have 1 h long and 360000 samples of each ECG data record. The sample of each ECG record was partitioned into 16 windows. Each window contains thousands of samples. Then DWT (db2 level 3) was performed in these subsamples. Total 20 coefficients per sample window are obtained. Total 20 * 16 = 320 attributes are obtained from 16 windows. After then we have applied Principal component analysis (PCA) for dimensionality reduction.

4 Principal Component Analysis and Its Method

Principal component analysis (PCA) is a statistical procedure that uses orthogonal transformation to convert a set of observation of possibly correlated variables into a set of linearly uncorrelated variables. PCA uses for dimensionality reduction of large data sets. Dimensionality reduction is a feature extraction technique. It represents the data using fewer variables. There are some other common applications of PCA denoising signals, blind source separation, and data compression. PCA transforms the data into a set of uncorrelated feature vectors called the principal components [8].

In this work PCA used for data reduction. The features vector which has extracted from the Apnea ECG has dimensionally reduced by PCA to identify the normal and abnormal ECG data. The covariance matrix, Eigen values and eigenvectors are computed. The Eigen values which are obtained are sorted along with their corresponding eigenvectors.

5   Classification

A neural network is used as the classifier so that it can be trained to understand the difference between features belonging to the different datasets or classes. In this work Multilayer Perceptron (MLP) neural network is used as a classifier to classify different subjects.

Multilayer consists of several layers and a feed forward structure with an error based training mechanism. The MLP and many other neural networks use a learning algorithm called back propagation, the input data is continually presented to the neural network, in each iteration an error is calculated by comparing the obtained output of the neural network to the desired output. This error is then feedback (back propagated) to the neural network and is used to modify the weights in such a way that the error decreases repetitively with each iteration and the neural model output approaches the desired outcome. This process is known as "training". The input layer in MLP consists of the extracted features, one or more hidden layers, and an output layer (which determines the class). Each layer in MLP consists of at least one neuron. From the input layer, an applied input passes the network in a forward direction through all layers [7] (Fig. 6).

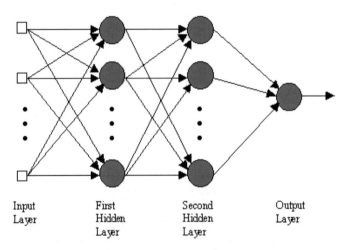

Input	First	Second	Output
Layer	Hidden	Hidden	Layer
	Layer	Layer	

Fig. 6. Architecture of MLP with two hidden layers

6   Simulation Results and Discussion

To verify the identity of different database, we used ECG signals derived from MITBH Arrhythmias database and Apnea ECG database [9].

Figure 7, the raw ECG signal is the input and filtered signal are obtained in the preprocessing stage to remove noise form the signal. After preprocessing of signals, the mentioned features were extracted from the ECG signals. Afterward, the extracted features were applied to Multilayer-Perceptron (MLP) classifier in order to verify different databases. In this study, we used 45 records from MIT-BIH Arrhythmias database and 70 records from Apnea ECG database. The result has been tested on different databases. Each database has separated into normal and abnormal class. Each record 10 s (3600 samples) from MIT-BIH Arrhythmia database and 1 h (360000 samples) long data from Apnea ECG database was the input. The target output was set to [0, 1], [1, 0] for normal and abnormal class. The test results are obtained by training the Multilayer Perceptron neural network using different numbers of hidden neurons in the hidden layer.

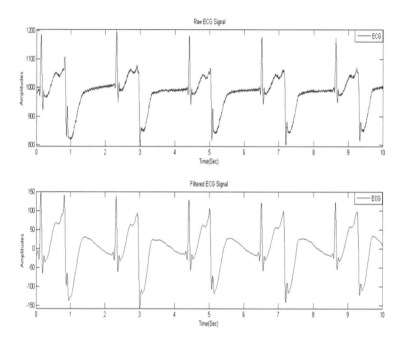

Fig. 7. Raw and filtered Arrhythmia ECG signal.

Accuracy of the proposed system is calculated using the following rule:
Accuracy% = TT + FF/(TF + FT + TT + FF) * 100.
Where;

  TT = Actual result is true and obtained result is also true.
  FF = Actual result is false and obtained result is also false.
  TF = Actual result is true and obtained result is false.
  FT = Actual result is false and obtained result is true.

Table 1 shows the results which are obtained using MLP classifier to recognize normal and abnormal ECG signals from different database. We used 11 hidden neurons for MIT-BIH Arrhythmias database. 'Trainlm' training function are used to updates weight and bias values with sigmoid activation function. From 45 records, 27 records are chosen for training and remaining 18 records are used for testing. We can see the best performance achieved in MIT-BIH Arrhythmia database 90% normal and 100% abnormal detection with 94.44% accuracy achieved. 1 record is misclassified as abnormal class.

Table 1. Simulation result of MIT-BIH Arrhythmias database

Database	No. ECG records		Accuracy (%)	
MIT-BIH Arrhythmia	Train	Test	Acc.	Overall
Normal	15	10	90%	94.44%
Abnormal	12	8	100%	

The Apnea ECG database is divided into two separate classes that are normal and abnormal. Each record of 1 h recording was picked up. Total 70 recordings are considered in this study. Total 90 numbers of DWT based features are splited into two different classes, after the dimensionality reduction using Principal Component Analysis (PCA) method. 15 numbers of hidden neurons are used in the back propagation neural network. Training data used for preparing the network architecture. For training and testing 70 ECG data are used. 48 ECG data are randomly separated for training and remaining used for validation and testing. Network are trained with scaled conjugate gradient propagation (trainscg). Sigmoid activation function are used. The output is determined in the form of 0 and 1. 0 for normal ECG (non-apnea) and 1 for abnormal ECG (apnea). Figure 8, provide the results such as overall performance metrics. Classification results are shown and given by a confusion matrix. According to the confusion matrix, 1 normal subject is misclassified as abnormal class (24 normal subjects are classified correctly out of 25) whereas 43 subjects are classified correctly as abnormal and 2 subjects are misclassified as normal. From this confusion matrix the classification accuracy for normal and abnormal class data was 96% and 95.6% respectively. The overall system performance was achieved with 95.7% accuracy using back propagation neural network.

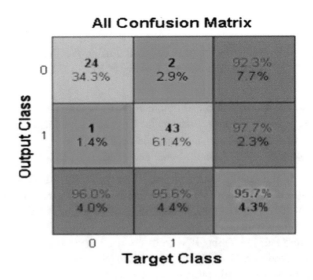

Fig. 8. Confusion matrix (classification result of Apnea ECG database)

We performed a comparison with other works based on various records and approach of Arrhythmia and Apnea–ECG database. Table 2 represent the comparison of our work and previous work, which shows that Neural Network gives better classification rate than other classifier for the detection of normal and abnormal ECG signal.

Table 2. Comparison table

Method	Database	Results
Pattern Recognition Tool (using 10 s complete ECG signal) [10]	Physiobank (MITBIH Arrhythmia)	100% Normal 84.6% Abnormal
Adaptive NeuroFuzzy interface system [11]	MIT-BIH Normal Sinus Rhythm & MIT-BIH Super ventricular Arrhythmia	100% Normal 91.48% Abnormal
ELM classifier SVM classifier BPN classifier [12]	MIT-BIH Arrhythmia	97% 73% (Accuracy) 64%
W-ECG features + PCA [13]	Apnea ECG	93.91% Accuracy
Sequency ordered Hadamard transform feature [14]	12 lead ECG signal (Apnea)	95.6% Accuracy
ECG signal features & SVM classifier [15]	Apnea ECG	96.5% accuracy
Measure of minutes of sleep disordered respiration [16]	Apnea ECG	91% accuracy
	MIT-BIH Arrhythmia	90% normal 100% Abnormal
Proposed work		(94.44% Accuracy)
	Apnea ECG	95.7% Accuracy

7 Conclusions

In these study, we have experimented with two different databases taken from physionet.org (i) normal and arrhythmia and (ii) normal and apnea – ECG signals. The total input samples of the signals are divided into windows of each class from different databases. The DWT was applied to the ECG signal in each window and extracted statistical information from the wavelet coefficients. PCA was only used for the detection of normal and abnormal ECG signals of Apnea ECG database for feature dimensionality reduction. Total 45 and 70 recordings are used from MITH-BIH Arrhythmia and the Apnea-ECG Database respectively for training and testing phase. In the training phase, the training data or the extracted features were trained through Multilayer Perceptron Neural Network with back propagation. The simulation results show the classification of different database of normal and abnormal ECG signal achieves 90% and 100% respectively with 94.4% accuracy from MIT-BIH Arrhythmia database with 11 hidden neurons whereas 96% normal and 95.6% abnormal detection with 95.7% accuracy from Apnea-ECG database using 15 numbers of hidden neurons in the hidden layer.

References

1. Mukhopadhay, S., Roy, A.B., Dey, N.: Wavelet based QRS complex detection of ECG signal. Int. J. Eng. Res. Appl. (IJERA) **2**, 2361–2365 (2012)
2. Sarkaleh, M.K., Shahbahram, A.: Classification of ECG Arrhythmias using discrete wavelet transform & neural network. Int. J. Comput. Sci. Eng. Appl. (IJCESA) **2**(1), 1 (2012)
3. http://www.medicinet.com/arrhythmia_irregular_heartbeat
4. Ain, S.N., Azam, M., Zainal, N.I., Sidek, K.A.: Development cardioid based graph ECG heart abnormalities technique. ARPN J. Eng. Appl. Sci. **10**, 9759–9765 (2015)
5. http://www.webmd.com/sleep_disorder/sleep-apnea
6. Almazaydeh, L., Elleithy, K., Faezipour, M.: Detection of obstructive sleep Apnea through ECG Signal Features. University of Bridgeport, Bridgeport, CT 06604, USA
7. Iqbal, F.-t.-Z., Sidek, K.A.: Cardioid graph ECG biometric using compressed QRS complex. In: International Conference on Bio Signal Analysis Processing and Systems (ICBAPS) (2015)
8. Rama, V., Rama Rao, C.B., Duggirala, N.: Analysis of signal processing techniques to identify cardiac disorders. Int. J. Innov. Res. Electr. Electron. Instrum. Control Eng. **3**(6) (2015)
9. https://physionet.org
10. Sharma, A., Bhardwaj, K.: Identification of normal and abnormal ECG using neural network. Int. J. Inf. Res. Rev. **2**, 695–700 (2015)
11. Sahu, N.K., Ayub, S., Saini, J.P.: Detection of normal ECG and Arrhythmia using adaptive neuro-fuzzy interface system. Int. J. Adv. Res. Comput. Sci. Softw. Eng
12. Subbiah, S., Patro, R., Subbthai, P.: Feature extraction and classification for ECG signal processing based on artificial neural network and machine learning approach. In: International Conference on Inter Disciplinary Research in Engineering and Technology (2015)
13. Rachim, V.P., Li, G., Chung, W.-Y.: Sleep Apnea classification using ECG signal wavelet-PCA features. Bio-Med. Mater. Eng. **24**, 2875–2882 (2014)
14. Kora, P., Annavarapu, A., Yadlapalli, P., Katragadda, N.: Classification of sleep Apnea using ECG-signal sequence ordered Hadamard Transform Features. Int. J. Comput. Appl. (0975–8887) **156**(14), 7–11 (2016)
15. Almazaydeh, L., Elleithy, K., Faezipour, M.: Obstructive sleep apnea detection using SVM-based classification of ECG signal features. In: 34th Annual International Conference of the IEEE EMBS San Diego, California USA, 28 August–1 September 2012

16. Chazal, P., Penzel, T., Heneghan, C.: Automated detection of obstructive sleep apnoea at different time scales using the electrocardiogram. Inst. Phys. Publ. **25**(4), 967–983 (2004)
17. http://en.wikipedia.org/wiki/Discrete_wavelet_transform
18. https://physionet.org/physiobank/database/apneaecg/additional-information.txt
19. http://mathworks.com

Dynamic Relation-Based Analysis of Objective Interestingness Measures in Association Rules Mining

Rachasak Somyanonthanakul[1(✉)] and Thannaruk Theeramunkong[1,2]

[1] School of Information, Computer, and Communication Technology,
Sirindhorn International Institute of Technology, Thammasat University,
Bangkok, Thailand
rachasaks@hotmail.com, thanaruk@siit.tu.ac.th
[2] The Royal Society of Thailand, Bangkok, Thailand

Abstract. While a large number of objective interestingness measures have been proposed to extract interesting rules from a dataset, most of them have been tested on a limited number of datasets that may not cover all possible patterns. This paper presents a framework to investigate relation among twenty-one interestingness measures on synthesized patterns (A → B), using all combinations of the six probabilities P(A, B), P(A, ¬B), P(¬A, B), P(¬A, ¬B), P(A) and, P(B) with a fixed number of occurrences. The partial order of interestingness measures is compared to that of another measure in order to characterize their similarity. The result shows 75 interrelation patterns of probabilities. An association rule mining is used to analyzed to describe for understanding their common and distinct properties.

Keywords: Association rules · Interestingness measurement
Synthesized patterns · Trend analysis

1 Introduction

Association rule mining [1] is often used to analyze interesting patterns, usually in a form of production rules, hidden in a transactional database [2]. To extract a strong rule, it is necessary to define a suitable measure of interestingness. So far, a large number of interestingness measures [3] have been invented to express various types of information, used for pruning trivial rules and/or ranking rules during the mining process. Since these measures have their own strength and weakness, they are used in different scenarios. In the past, several works [4] were conducted to analyze characteristics of different measures, including similarity among measures, mathematics properties etc. However, these works focuses on a limited number of focused cases; in terms of a 2 × 2 contingency table $n(A, B)$, $n(A, ¬B)$, $n(¬A, B)$ and $n(¬A, ¬B)$ or a set of realworld datasets, which are not guaranteed to cover all varieties in terms of possible situations, or sometimes may include unbalances. As a solution, this work is proposed to use a synthetic patterns (A → B) of (which enumerates all feasible combinations of $n(A, B)$, $n(A, ¬B)$, $n(¬A, B)$ and, $n(¬A, ¬B)$ where the total number of occurrences are

© Springer Nature Switzerland AG 2019
T. Theeramunkong et al. (Eds.): iSAI-NLP 2017, AISC 807, pp. 38–46, 2019.
https://doi.org/10.1007/978-3-319-94703-7_4

fixed, in order to analyze the measures. By this pattern, we investigate correlation analysis, similarity and dissimilarity characteristics of the measures in terms of qualitative and quantitative points of view [5]. The outline of this paper is as follows. The rest of this paper is organized as follows. In Sect. 2, we review interesting interestingness. In Sect. 3, we describe the analytic methods. In Sect. 4, we show the result of experiments. In Sect. 5, we discussion and conclusion of this paper.

2 Interestingness and Its Related Work

2.1 Pairwise Association Rule

In contrast, the subjective measures where human factor and domain are involved. The objective measures are rooted on a data-driven approach, can be used for evaluating the quality of association patterns, independently of domain with minimal labor from the users. Theoretically, an objective measure requires a threshold to filter quality patterns. Most objective measures are usually computed based on frequency counts tabulated in a contingency table as shown in Table 1.

Table 1. A two-way contingency table for variable A and B

	B	$\bar{B}$	$B + \bar{B}$
A	$f.$	$f_J = fi. - f.$	f_1
$\bar{A}$	$f_{/.} = f_1. - f..$	$f_{//} = N-f. - f._1 - f_1$	$f_{/1} = f_{/.} - f_{//}$
$A + \bar{A}$	$f_1.$	$f_{1/} = f_{/} + f_{//}$	N

In decades, several different interestingness measures have been developed for representing the strength of association rules. They are various in purpose and usefulness. Traditionally, three most conventional measures are support, confidence and lift (interestingness). Suppose that the association rule considered is $A \rightarrow B$. The support probability of A $(P(A))$. indicates how many transactions the association rule holds. Two alternative forms of supports are the absolute value and the ratio to the total number of transactions. The confidence $(P(B|A))$ indicates the correlation among A and B, that is when an event A occurs, how likely the other event B will occur. Similar to the confidence, the lift shows the ratio of posterior probability of A when B is known $(P(A|B))$ to the prior. However, by defining different association measures, it is possible to find different types of association patterns or rules that are appropriate for different types of data and applications. This situation is analogous to that of using different objective functions for measuring the goodness of a set of clusters in order to obtain different types of clustering.

2.2 Previous Works on Measurement Analysis

The works were studied on the usage and comparison of objective interestingness in data mining. [6] identified the difference between objective and subjective interestingness. An objective measure depends only on the raw dataset.

A subjective measure responsible between the data and the user of these data. [7, 8] selected 10 datasets and 20 measurements to cluster observe differences depending. The results shown that some measures are monotonically property of the confidence, while others are monotonically increasing transformations of the lift so that such measures will rank the rules according to the same order. [9] was the first study in the clustering measurement based on ranking behavior using 35 measurements with 2 datasets. They selected datasets that consists of different properties. [10] focused on 38 interestingness measures that perform on a real application in the medical data. That interestingness measures were mainly used to remove meaningless rules rather than to discover really interesting ones for a human user, since they do not include domain knowledge. [11] studied a theoretical framework that uses both set theory and a probabilistic point of view to analyze interestingness measures. Their work described 16 measures, which they categorized based on the distinction. [12] consider 38 interestingness measures in their design of a methodology for pruning association rules. In particular, they show how each interestingness measure performs in terms of coverage after poor rules have been removed versus when all rules are included. Unfortunately, it is difficult to infer anything from these results of the relative similarity or correlation among interestingness measures. Table 2 illustrates a summary description for 21 common measures [4].

Table 2. Interestingness measurements.

Interestingness	Definition
ϕ-coefficient	$\dfrac{P(A,B)-P(A)P(B)}{\sqrt{P(A)P(B)(1-P(A))(1-P(B))}}$
Goodman-Kruskal's	$\dfrac{\sum_j max_k P(A_j,B_k) + \sum_k max_j P(A_j,B_k) - max_j P(A_j) - max_k P(B_k)}{2 - max_j P(A_j) - max_k P(B_k)}$
Odds ratio	$\dfrac{P(A,B)P(\bar{A},\bar{B})}{P(A,\bar{B})P(\bar{A},B)}$
Yule's Q	$\dfrac{P(A,B)P(\overline{AB})-P(A,\bar{B})P(\bar{A},B)}{P(A,B)P(\overline{AB})+P(A,\bar{B})P(\bar{A},B)} = \dfrac{\alpha-1}{\alpha+1}$
Yule's Y	$\dfrac{\sqrt{P(A,B)P(\overline{AB})}-\sqrt{P(A,\bar{B})P(\bar{A},B)}}{\sqrt{P(A,B)P(\overline{AB})}+\sqrt{P(A,\bar{B})P(\bar{A},B)}} = \dfrac{\sqrt{\alpha}-1}{\sqrt{\alpha}+1}$
Kappa	$\dfrac{P(A,B)+P(\bar{A},\bar{B})-P(A)P(B)-P(\bar{A})P(\bar{B})}{1-P(A)P(B)-P(\bar{A})P(\bar{B})}$
Mutual Information	$\dfrac{\sum_i \sum_j P(A_i,B_j)\log\frac{P(A_i,B_j)}{P(A_i)P(B_j)}}{min\left(-\sum_i P(A_i)\log P(A_i), \sum_j P(B_j)\log P(B_j)\right)}$
J-measure	$max\left(P(A,B)\log\left(\frac{P(B|A)}{P(B)}\right)+P(A\bar{B})\log\left(\frac{P(\bar{B}|A)}{P(\bar{B})}\right), P(A,B)\log\left(\frac{P(A|B)}{P(A)}\right)+P(\bar{A},B)\log\left(\frac{P(\bar{A}|B)}{P(A)}\right)\right)$
Gini index	$max\left(\begin{array}{l}(P(A)[P(B|A)^2+P(\bar{B}|A)^2]+P(\bar{A})[P(B|\bar{A})^2+P(\bar{B}|\bar{A})^2]) - P(B)^2-P(\bar{B})^2 \\ ,(P(B)[P(A|B)^2+P(\bar{A}|B)^2]+P(\bar{B})[P(A|\bar{B})^2+P(\bar{A}|\bar{B})^2]) - P(A)^2-P(\bar{A})^2\end{array}\right)$

(continued)

Table 2. (*continued*)

Interestingness	Definition
Support	$P(A, B)$
Confidence	$\max(P(B|A), P(A|B))$
Laplace	$\max\left(\frac{NP(A,B)+1}{NP(A)+2}, \frac{NP(A,B)+1}{NP(B)+2}\right)$
Conviction	$\max\left(\frac{P(A)P(\bar{B})}{P(A\bar{B})}, \frac{P(B)P(\bar{A})}{P(BA)}\right)$
Interest	$\frac{P(A,B)}{P(A)P(B)}$
Cosine	$\frac{P(A,B)}{\sqrt{P(A)P(B)}}$
Piatetsky-Shapio's	$P(A, B) - P(A)P(B)$
Certainty factor	$\max\left(\frac{P(B|A)-P(B)}{1-P(B)}, \frac{P(A|B)-P(A)}{1-P(A)}\right)$
Added Value	$\max(P(B|A) - P(B), P(A|B) - P(A))$
Collective strength	$\frac{P(A,B) + P(\overline{AB})}{P(A)P(B) + P(\bar{A})P(\bar{B})} \times \frac{1-P(A)P(B)-P(\bar{A})P(\bar{B})}{1-P(A,B)-P(\overline{AB})}$
Jaccard	$\frac{P(A,B)}{P(A) + P(B) - P(A,B)}$
Klosgen	$\sqrt{P(A, B)}\max(P(B|A) - P(B), P(A|B) - P(A))$

3 Analytical Methods

3.1 Variable Probabilities

In this section, we conduct a synthetic pattern $(A \rightarrow B)$ to recognize the pattern and tendency of the six probabilitie; $p(A,B), p(A, \neg B), p(\neg A, B), p(\neg A, \neg B)$ and $p(A)$ and $p(B)$. In our experiment, we choose three main independent variables; $p(A, \neg B), p(\neg A, B), p(\neg A, \neg B)$ and the others three dependence variables;. Table 4 shows contingency table where $f \in \{100k | k \in \{1, 2, \ldots, 10\}\}, f_{1+} \in \{100k | k \in \{1, 2, \ldots, 10\}\}, f_{11} < f_{1+}, f_{11} < f_{+\cdot}$, and $N = 1,000$

3.2 Dynamic-Tendency Pattern

To extend previous work [11], the twenty-seven tendency patterns was proposed. This work, we expand its to cover all possible combination using the pairwise comparison method. We give a fix total number N, three main free variables, p(A), p(B)., and $p(A, B)$, f_{-1}, f_1 and $f_\cdot$ are varied for investigation, each with three possible stages. We compare all synthesis patterns with each other. In the experiment, we apply the pairwise comparison technique to compare each candidate table matched a paired record with each of the other alternative table. Each candidate table gets positive trend (P) for a pair record least than, negative trend (N) for great than and equal trend (E) for a tie.

3.3   Correlation Analysis Using Dynamic-Tendency Pattern

We compare twenty-one measurement over the entire set of rules generated by association rule mining. An analysis approach performs association rule mining and consider only the top N rules it generates. An association rule generates rules that match the minimum support and confidence. Therefore, considering the top 100 rules to take advantage to some of the interestingness measures, namely confidence and its derivatives. We run the standard the Associate function in Weka [14] with a minimum confidence threshold of 0.9, set the minimum support threshold low enough for each dataset to generate at least 100 rules, and select only 100 rules from the result set.

4   Measurement Analysis Results

4.1   Structure of Variable Probabilities

We generated 286 feasible two-way contingency tables to conduct our experiment. Table 3 shows two synthetic patterns that contained 2×2 contingency tables

The synthetic pattern is composed of a number of occurrences that described by a number of variables. We set all integer number satisfying.. $< f_{.1}, f_{..} < f_{1.}$, and $N = 1000$ and increase 100 in all variables for every step increase. The variable. $_1$ and f_1 are fixed 500 and $f_{..}$ increases from 300 to 400. The comparison results show E (equal) in variable $f_{.1}$ and $f_{1.}$. The variable $f_{.}$ show P (positive) of pair #1 vs. #2 and N (negative) of pair #2 vs. #1.

In this step, we apply the pairwise comparison method for twenty-one interestingness measures in every pattern (81,550 patterns). The interestingness measurement was compared to show tendency in each pair. The pattern of all records was measured using twenty-one interestingness measures. The result of comparison give positive trend (P) for a pair record least than, negative trend (N) for great than and equal trend (E) for a tie.

Table 3. The sample of variable probability

No.	f_{1+}	f_{+1}	f_{11}	f_{00}	f_{01}	f_{10}	f_{0+}	f_{+0}	N
#1	500	500	**300**	300	200	200	500	500	1000
#2	500	500	**400**	400	100	100	500	500	1000
#1 vs. #2	E	E	P	P	N	N	E	E	E
#2 vs #1	E	E	N	N	P	P	E	E	E

4.2   Dynamic-Tendency Pattern Analysis

In this section, we extend previous work [11] from the twenty-seven tendency patterns to seventy-five pattern. We use structure of variable probabilities in Sect. 4.1 to compare with each all over pattern. In the experiment, we apply the pairwise comparison method to compare each candidate table matched a paired record with each of the other alternative table. Each candidate table gets positive trend (*P*) for a pair record least than, negative trend (*N*) for great than and equal trend (*E*) for a tie.

We generate 81,550 patterns using pairwise comparison method. Then, seventy-five pattern [30] was used to be filter and count number of each pattern. Table 6 shows the all possible pattern in Table 6.

Table 4. Total number of tendency pattern

#	.1	1.	.	//	/.	./	#	.1	1.	.	//	/.	./
1*	E	P	E	N	P	E	38*	E	N	N	E	E	P
2*	E	N	E	P	N	E	39	E	P	P	N	P	N
3*	P	E	E	N	E	P	40	E	N	N	P	N	P
4*	N	E	E	P	E	N	41	P	P	P	N	E	N
5*	P	P	E	N	P	P	42	N	N	N	P	E	P
6*	N	N	E	P	N	N	43	P	P	P	N	P	N
7*	P	P	P	N	E	E	44*	N	N	N	P	N	P
8*	N	N	N	P	E	E	45*	E	E	P	P	N	N
9	P	P	P	N	P	E	46*	E	E	N	N	P	P
10	N	N	N	P	N	E	47	P	P	P	E	N	N
11	P	P	P	N	E	P	48	N	N	N	E	P	P
12	N	N	N	P	E	N	49	P	P	P	N	N	N
13	P	P	P	N	P	P	50	N	N	N	P	P	P
14	N	N	N	P	N	N	51*	E	P	N	N	P	P
15*	P	N	E	E	N	P	52*	E	N	P	P	N	N
16	N	P	E	E	P	N	53*	P	N	N	N	E	P
17*	P	E	P	E	N	E	54*	N	P	P	P	E	N
18*	N	E	N	E	P	E	55	P	E	N	N	P	P
19	P	N	E	N	N	P	56*	N	E	P	P	N	N
20	N	P	E	P	P	N	57*	P	P	N	N	P	P
21	P	E	P	N	N	P	58*	N	N	P	P	N	N
22	N	E	N	P	P	N	59	P	E	P	P	N	N
23	P	P	P	N	N	E	60	N	E	N	N	P	P
24	N	N	N	P	P	E	61	P	N	N	E	N	P
25	P	P	P	N	N	P	62*	N	P	P	E	P	N
26	N	N	N	P	P	N	63	P	N	N	N	N	P
27	P	N	E	P	N	P	64	N	P	P	P	P	N
28*	N	P	E	N	P	N	65	P	N	P	P	N	N
29	P	N	P	P	N	E	66	N	P	N	N	P	P
30*	N	P	N	N	P	E	67	P	N	N	P	N	P
31*	P	N	P	E	N	P	68	N	P	P	N	P	N
32	N	P	N	E	P	N	69*	E	P	P	P	N	N
33	P	N	P	N	N	P	70	E	N	N	N	P	P
34	N	P	N	P	P	N	71	P	P	P	P	N	N
35	P	N	P	P	N	P	72	N	N	N	N	P	P
36	N	P	N	N	P	N	73	P	N	N	N	P	P
37	E	P	P	E	E	N	74	N	P	P	P	N	N
							75	E	E	E	E	E	E

Table 5. The correlation Association rules (minimum support = 0.4)

No.	LHS		RHS	conf	lift	lev	conv
1	Q=N	→	Odd Ratio=N	1	2.24	0.22	18285.11
2	Q=P	→	Odd Ratio=P	1	2.24	0.22	18285.11
3	Q=N	→	Y=N	1	2.29	0.23	18643.75
4	Q=P	→	Y=P	1	2.29	0.23	18643.75
5	Q=N, Y=N	→	Odd Ratio=N	1	2.24	0.22	18285.71
6	Odd Ratio=N, Q=N	→	Y=N	1	2.29	0.23	18643.75
7	Q=N	→	Odd Ratio=N, Y=N	1	2.31	0.23	18715.92
8	Q=P, Y=P	→	Odd Ratio=P	1	2.24	0.22	18285.71
9	Odd Ratio=P, Q=P	→	Y=P	1	2.29	0.23	18643.75
10	Q=P	→	Odd Ratio=P, Y=P	1	2.31	0.23	18715.92
11	Coefficient=N, Conviction=N, Piatetsky Shapiro=N, Added Value=N, CollectiveStrength=N	→	Kappa=N	1	2.07	0.21	233.9
12	Coefficient=P, Conviction=P, Piatetsky Shapiro=P, Added Value=P, Collective Strength=P	→	Kappa=P	1	2.07	0.21	233.9
13	Coefficient=N, Conviction=N, Piatetsky Shapiro=N, Added Value=N	→	Kappa=N	1	2.07	0.21	219.8
14	Coefficient=P, Conviction=P, Piatetsky-Shapiro=P, AddedValue=P	→	Kappa=N	1	2.07	0.21	219.8

Table 6. The correlation of interestingness measurement

Group	Measurement
1	Yule's Q, Yule's Y, Odds ratio
2	f-coefficient, Laplace, Cosine, Certainty factor, Collective strength, Kappa, Conviction, Piatetsky-Shapio's
3	Interest, Jaccard
4	Mutual Information, Gini index, J-measure

4.3 Measurement Correlation Analysis

In this section, we analyze the correlation of measure using association method. The trend analysis in Sect. 4.2 is extracted the correlation among measurement. We use WEKA software in scheme of the association rules mining. In this section, we use total a number of tendency interestingness of 81,510 instances, 30 instances. The threshold was set the minimum support of 0.4, minimum confidence of 0.9.

We select top 1,000 rules to find association of tendency. The rules show left-hand site (LHS) and right-hand site (RGH). Table 7 show the samples rules. In the first rule, $(Q = N) \rightarrow (Odd\ Ratio = N)$ is measure Yule's Q = negative associated with Odd Ration = negative. Four measures are *conf* (confident) is an estimation of conditioned probability, *lift* is negatively correlated, if the value is less than 1 otherwise, positively correlated, *lev* (leverage) is the proportion of additional elements covered by both the premise and consequence above the expected if independent., and *conv* (conviction) is the number of instances for which it predicts correct.

Table 5 shows the group of correlations of interestingness measurement. We select the four group of correlations. A group of measurements are characterized following density based clustering. The first group are strongly occurrence that consists of three measurement: *Yule's Q, Yule's Y,* and *Odds ratio.* The second group are eight measurements that consists of f-*coefficient, Laplace, Cosine, Certainty factor, Collective strength, Kappa, Conviction, Piatetsky-Shapio's.* Third group consists of *Interest* and *Jaccard* and the last group is *Mutual Information, Gini index, J-measure.*

5 Discussion and Conclusion

This paper presents a framework to investigate relation among twenty-one interestingness measures on synthesized patterns (A → B), using the partial order of a measure. The first step, we generated the variable probabilities that is the same method of [3, 5, 6]. In this step, the synthetic pattern was constructed covering all possible patterns. The coverage pattern (A → B) are not mention in [7–9].

The second phase, dynamic-tendency pattern analysis is proposed using pairwise method that is mention in [3]. However, the result shows seventy-five pattern that cover all possible patterns (A → B) and not show in [3, 5–8]. The all patterns can be reference in datamining.

In the last section, The WEKA software is used to analysis the correlation of interestingness measurement. The four main groups of measurements are presented to be measurement correlation of dynamic pattern. The mining threshold are set in default values and present in limited top rules. the future work, complex pattern will generate in order to investigate relationship between objective interestingness measure and association pattern.

Acknowledgements. This work has been supported by Sirindhorn International Institute of Technology, Thammasat University.

References

1. Agrawal, R., Imielinski, T., Swami, A.: Mining association rules between sets of items in large databases. In: ACM SIGMOD International Conference on Management of Data, Washington DC, pp. 207–216 (1993)
2. Agrawal, R., Srikant, R.: Fast algorithms for mining association rules. In: Proceedings of the 20th International Conference on Very Large Data Bases, pp. 487–499 (1994)
3. Geng, L., Hamilton, H.J.: Interestingness measures for data mining: a survey. ACM Comput. Surv. (CSUR) **38**(3), 9 (2006)
4. Tan, P.N., Kumar, V., Srivastava, J.: Selecting the right objective measure for association analysis. Inf. Syst. **29**(4), 293–313 (2004)
5. Tew, C., Giraud-Carrier, C., Tanner, K., Burton, S.: Behavior-based clustering and analysis of interestingness measures for association rule mining. Data Min. Knowl. Disc. **28**(4), 1004–1045 (2014)

6. Tan, P.N., Kumar, V., Srivastava, J.: Selecting the right interestingness measure for association patterns. In: The Eighth ACM SIGKDD International Conference on Knowledge Discovery and Data Mining, Edmonton, Alberta, pp. 32–41 (2002)
7. McGarry, K.: A survey of interestingness measures for knowledge discovery. Knowl. Eng. **20**(1), 39–61 (2005)
8. Vaillant, B., Lenca, P., Lallich, S.: A clustering of interestingness measures. In: Proceedings of the 7th international conference on discovery science (LNAI 3245), pp 290–297 (2004)
9. Huynh, X.H., Guillet. F., Briand, H.: Discovering the stable clusters between interestingness measures. In: Proceedings of the 8th International Conference on Enterprise Information Systems: Databases and Information Systems Integration, pp 196–201 (2006)
10. Ohsaki, M., Kitaguchi, S., Okamoto, K., Yokoi, H., Yamaguchi, T.: Evaluation of rule interestingness measures with a clinical dataset on hepatitis. In: Proceedings of the 8th European Conference on Principles and Practice of Knowledge Discovery in Databases (LNAI 3203), pp 362–373 (2004)
11. Somyanonthanakul, R., Theeramunkong, T.: An Investigation of Objective Interestingness Measures for Association Rule Mining. In: Pacific Rim International Conference on Artificial Intelligence. pp. 472–481. Springer International Publishing (2016)
12. Lenca, P., Meyer, P., Vaillant, B., Lallich, S.: On selecting interestingness measures for association rules: user oriented description and multiple criteria decision aid. Eur. J. Oper. Res. **184**(2), 610–626 (2008)
13. Yao, Y., Zhong, N.: An analysis of quantitative measures associated with rules. In: Proceedings of the 3rd Pacific-Asia Conference on Knowledge Discovery and Data Mining. LNCS, vol. 1574, pp 479–488 (1999)
14. Kannan, S., Bhaskaran, R.: Association rule pruning based on interestingness measures with clustering. Int. J. Comput. Sci. **6**(1), 35–45 (2009)
15. Witten, I., Eibe, F.: Data mining: practical machine learning tools with Java

Image Encryption Using Cellular Neural Network and Matrix Transformation

Gangyi Hu[1](✉), Jian Qu[2], and Sumeth Yuenyong[2]

[1] College of Big Data and Intelligence Engineering, Southwest Forestry University, Bailong Road, Panlong District, Kunming 650024, Yunnan, China
hugangyi604@126.com
[2] School of Science and Technology, Shinawatra University, 99 Moo 10 Bang Toey, Sam Khok District, Pathum Thani, 12160, Thailand
{qu.j,sumeth.y}@siu.ac.th

Abstract. This paper proposes an image encryption algorithm based on CNN (Cellular Neural Network) chaotic system and matrix transformation. The algorithm uses the initial State of CNN as the encryption key, which generates five-dimensional chaotic sequence. Then the image pixel values were changed by performing XOR operation between the original image pixel values and the modified chaotic sequence. Finally, the pixel positions were changed using a construction matrix, resulting in the cipher image. The experiment results show that this algorithm has good encryption effect, strong key sensitivity and high security.

Keywords: Image encryption · Cellular Neural Network · Chaotic system

1 Introduction

Traditional image encryption methods such as the Arnold transformation [1] have limitations in cipher image with low information entropy and low number of pixel change ratio. Thus, their security can be improved. Chaos has the characteristics of randomness and uncorrelatedness, these properties are excellent for satisfying the image encryption requirements. It has been proven in other works that using chaotic systems may obtain strong security [2–5].

The cellular neural network (CNN) has the characteristics of parallel signal processing and be able to generate complex chaotic behavior. Therefore, it plays an important role in the image encryption field [6].

This paper proposes an image encryption algorithm based on CNN five dimensional chaotic system and a matrix transformation, which we call W. We give the coefficient parameters of the CNN system that ensures chaotic property. Then the five initial state values and one step size value were fed into the CNN system to get a chaotic sequence. Thus, these six values are the encryption key.

During the process of encrypting an image, XOR operation between those modified sequences and the original image pixel value is performed. After that the pixel positions are changed according to the W matrix transformation to get the cipher image. The

© Springer Nature Switzerland AG 2019
T. Theeramunkong et al. (Eds.): iSAI-NLP 2017, AISC 807, pp. 47–57, 2019.
https://doi.org/10.1007/978-3-319-94703-7_5

proposed algorithm has the advantages of high information entropy, strong sensitivity and high security.

2   The Cellular Neural Network Chaotic System

Cellular Neural Network was proposed by Chua et al., in 1988 [7]. It consists a number of circuit units which are called cells. Every cell C_{ij} has a state x_{ij}, an output y_{ij} and an input u_{ij}, the equivalent circuit of a cell is shown in Fig. 1.

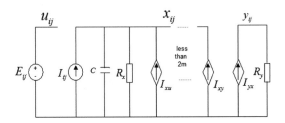

Fig. 1.   The equivalent circuit of a cell C_{ij}

The voltage controlled current sources have the features as shown in (1).

$$I_{xy} = Ay_{ij}\quad I_{xu} = Bu_{ij} \tag{1}$$

Where A and B are matrices that represent the weight of the feedback connections between the cell C_{ij} and the output y_{ij}; and the input weight between the cell C_{ij} and the input u_{ij}, respectively. These matrices are called "templates" in CNN terminology.

The equivalent equations for one cell C_{ij} can be obtained as shown in (2) [8, 9].

$$C\frac{dx_{ij}(t)}{dt} = -\frac{x_{ij}(t)}{R_x} + \sum_{k,l\in N_{ij}(r)} A_{kl}y_{kl}(t) + \sum_{k,l\in N_{ij}(r)} B_{kl}u_{kl}(t) + I_{ij} \tag{2}$$

And the output equation is in (3).

$$y_{ij}(t) = \frac{1}{2}(|x_{ij}(t) + 1| - |x_{ij}(t) - 1|) \tag{3}$$

Where, A is the feedback template matrix and B is the input template matrix, I_{ij} is the threshold matrix, C and R_x are the circuit parameters.

Assuming the circuit parameters $C = 1$, $R_x = 1$, the template matrix parameters take a specific group of values are shown in (4). These parameters are fixed to ensure chaotic property. However, different initial state values and step size value generates unique chaotic sequence [10]. Substituting (4) and (3) into (2), we can obtain (5), which is the update equation for our 5D chaotic CNN.

$$A = 0 \text{ except } a_{44} = 202; \ I = 0$$

$$B = \begin{bmatrix} 0 & 0 & -1 & -1 & 0 \\ 0 & 2 & 1 & 0 & 0 \\ 11 & -12 & 0 & 0 & 0 \\ 92 & 0 & 0 & -95 & -1 \\ 0 & 0 & 15 & 0 & -2 \end{bmatrix} \tag{4}$$

$$\begin{cases} \dfrac{dx_1}{dt} = -x_3 - x_4 \\[2mm] \dfrac{dx_2}{dt} = 2 * x_2 + x_3 \\[2mm] \dfrac{dx_3}{dt} = 11 * x_1 - 12 * x_2 \\[2mm] \dfrac{dx_4}{dt} = 92 * x_1 - 95 * x_4 - x_5 + 101 * (|x_4 + 1| - |x_4 - 1|) \\[2mm] \dfrac{dx_5}{dt} = 15 * x_3 - 2 * x_5 \end{cases} \tag{5}$$

The value of h and x_i ($i = 1, 2 \ldots 5$) can be set as any value of any digit length up to machine precision. As will be shown in the result section, encryption is highly sensitive to the key values. That is, if one of the key values was changed by only a tiny amount, the cipher image cannot be successfully decrypted. Thus the key space (the number of possible keys) is very large.

3   Construction of the W Matrix Transformation

According to Arnold matrix transformation, we construct a W transform matrix which can be used for changing the image pixel position. The expression of the W transformation matrix whose order is n(≥ 2) as shown in (6) [11]. In particular, the expressions for the second order and third order transformation of matrix W are showing in (7).

$$W_{n \times n} = \begin{pmatrix} 2 & 1 & 1 & 1 & \ldots & 1 & 1 \\ 3 & 1 & 2 & 2 & \ldots & 2 & 2 \\ 4 & 1 & 2 & 3 & \ldots & 3 & 3 \\ \ldots & \ldots & \ldots & \ldots & \ldots & \ldots & \ldots \\ n & 1 & 2 & 3 & \ldots & n-2 & n-1 \\ n+1 & 1 & 2 & 3 & \ldots & n-2 & n-1 \end{pmatrix} \tag{6}$$

$$W_{2 \times 2} = \begin{bmatrix} 2 & 1 \\ 3 & 1 \end{bmatrix} \quad W_{3 \times 3} = \begin{bmatrix} 2 & 1 & 1 \\ 3 & 1 & 2 \\ 4 & 1 & 2 \end{bmatrix} \tag{7}$$

The transformation of matrix W is periodic, the prove process is as follows.

Assume $|W|$ is the determinant of matrix W, we can get $|W_{2\times2}| = -1$ and $|W_{3\times3}| = 1$. According to the mathematical induction, assume the determinant value of k order matrix W is $(-1)^{k+1}$ as shown in (8), and the determinant value of $k + 1$ order matrix W is shown in (9).

$$|W_{k\times k}| = \begin{vmatrix} 2 & 1 & 1 & 1 & \dots & 1 & 1 \\ 3 & 1 & 2 & 2 & \dots & 2 & 2 \\ 4 & 1 & 2 & 3 & \dots & 3 & 3 \\ \dots & \dots & \dots & \dots & \dots & \dots & \dots \\ k & 1 & 2 & 3 & \dots & k-2 & k-1 \\ k+1 & 1 & 2 & 3 & \dots & k-2 & k-1 \end{vmatrix} = (-1)^{(k+1)} \tag{8}$$

$$|W_{(k+1)\times(k+1)}| = \begin{vmatrix} 2 & 1 & 1 & 1 & \dots & 1 & 1 \\ 3 & 1 & 2 & 2 & \dots & 2 & 2 \\ 4 & 1 & 2 & 3 & \dots & 3 & 3 \\ \dots & \dots & \dots & \dots & \dots & \dots & \dots \\ k+1 & 1 & 2 & 3 & \dots & k-1 & k \\ k+2 & 1 & 2 & 3 & \dots & k-1 & k \end{vmatrix} \tag{9}$$

Next, we subtract the last column of (9) from the first (which does not change the determinant). The result is shown in (10).

$$|W_{(k+1)\times(k+1)}| = \begin{vmatrix} 2 & 1 & 1 & 1 & \dots & 1 & 0 \\ 3 & 1 & 2 & 2 & \dots & 2 & 0 \\ 4 & 1 & 2 & 3 & \dots & 3 & 0 \\ \dots & \dots & \dots & \dots & \dots & \dots & \dots \\ k+1 & 1 & 2 & 3 & \dots & k-1 & 1 \\ k+2 & 1 & 2 & 3 & \dots & k-1 & 1 \end{vmatrix} \tag{10}$$

For the determinant of (10), we do the expansion through the column $k + 1$ which can get the result is

$$|W_{(k+1)\times(k+1)}| = |W_{k\times k}| + (-1)^{(k+2)} \times 2 \tag{11}$$

Follow the mathematical induction, substitute (8) in to (11), we can obtain (12).

$$|W_{(k+1)\times(k+1)}| = (-1)^{(k+1)} + (-1)^{(k+2)} \times 2 = (-1)^{(k+2)} \tag{12}$$

It can be known from the mathematical induction. For the all integer order matrix W, the determinant value of it is $|W_{n\times n}| = (-1)^{(n+1)}$.

From the value of the determinant above, the value of $|W|$ is coprime with N, which means the transformation from the matrix W such as $[X'] = W[X'] \bmod N$ has the characteristic of periodic. Because of this periodic. After lots of times transformation, it can go back to the original status. This matrix transformation is bijective.

The period of W matrix transformation is twice as much of the Arnold matrix transformation. This can increase the security of the encryption [11].

In this algorithm, we use the three order W matrix transformation for changing the pixel position. Firstly, it makes the image to three levels, and gets the row coordinate matrix and column coordinate matrix, respectively as shown in Fig. 2.

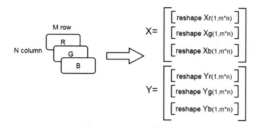

Fig. 2. The row and column coordinate matrix of an image. (size = 3, (row * column)).

For the row coordinate axis shuffle process, we change it by the product of W matrix and row pixel coordinate matrix as shown in (13). The column coordinates are changed as the same way showing in (14).

$$[X'_{i+1}] = W[X'_i] \bmod (M) \qquad X'_1 = X \tag{13}$$

$$[Y'_{i+1}] = W[Y'_i] \bmod (M) \qquad Y'_1 = Y \tag{14}$$

Where, X, Y represent the original coordinates of the image pixels, X', Y' represent the transformed coordinates of the image pixels, M represents the numbers of rows, N represents the number of columns. i is the numbers of iteration, which is less than the period of W matrix transformation.

4 The Proposed Algorithm

The process of this algorithm is to update the state of CNN from the initial state value and the step size in (5) to generate five-dimensional chaotic sequence. Then replace the image pixel values by performing XOR operation between the original image pixel values and the modified sequence to change the gray values in the image. Finally, the pixel position is changed according to the W matrix to obtain the cipher image.

4.1 Selection of the Key

In this algorithm, we select the initial state value x_i ($i = 1, ..., 5$) and step size h of the CNN chaotic system as the key for image encryption. Each key generates unique chaotic sequence and can be set to arbitrary values. This ensures the security of the key due to the large key space.

4.2 The Encryption Process

The step of this image encryption algorithm is as below.

1. Read the original image, and set the iteration initial value and step size of this CNN system.
2. Perform CNN iteration according to (5), which can get the five-dimensional chaotic sequence in (15).

$$\{x_1(k), x_2(k), x_3(k), x_4(k), x_5(k)|k = 0, 1, 2, ...\} \tag{15}$$

3. Because the chaotic sequence values are small decimal numbers on the order of 1, in order to be able to perform the XOR operation with the image pixel value, some transformations of the raw sequences are needed. We perform expand, union and modulo operation to all the value of this five-dimensional chaotic sequence to combine them into three 1-D chaotic sequences (one per each color channel). The value range of these three sequences is normalized from 0 to 255. The details of this process is given in the appendix. This process does not change the chaotic property of the CNN output sequence because any linear combination or elementary function of a chaotic sequence is still chaotic [12].
4. Perform XOR operation between the original image pixel values and these modified sequences to change the pixel values of the image.
5. Change the pixel position of the result from step 4 according to (13) and (14), which results in the final cipher image.

The process of image decryption is firstly using the encryption key as input for the CNN system to generate the same chaotic sequence as during the encryption. Change the pixel positions according to the W matrix transform to un-shuffle the image, and replace the image pixel values back by performing XOR operation again with the same modified chaotic sequence to recover the original image.

5 The Experimental Results

In this experiment, the Lena, Cameraman and Elephant images with the size of 256×256 have been used for testing images. The results are shown in Fig. 3. We can see clearly that the histograms of the cipher images are more uniform than that of the original images. It shows that the cipher images are randomly dispersed, which can hide the original images information.

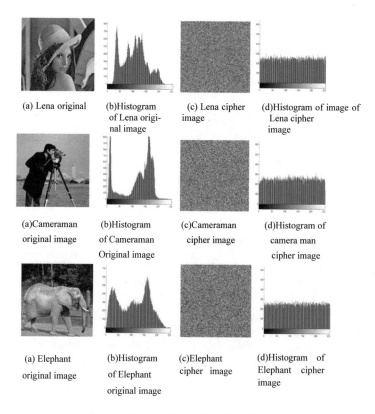

(a) Lena original | (b)Histogram of Lena original image | (c) Lena cipher image | (d)Histogram of image of Lena cipher image

(a)Cameraman original image | (b)Histogram of Cameraman Original image | (c)Cameraman cipher image | (d)Histogram of camera man cipher image

(a) Elephant original image | (b)Histogram of Elephant original image | (c)Elephant cipher image | (d)Histogram of Elephant cipher image

Fig. 3. The experimental results of the three images

5.1 NPCR Index Analysis

The number of pixel changing ratio (NPCR), is the ratio of the pixels in the cipher image that change when one pixel is the original image is changed.

In the simulation experiment, we get the NPCR value of the Lena, Cameraman and Elephant are 0.9959, 0.9960 and 0.9960, respectively. Meaning that even when one pixel in the original image is changed, the cipher image change completely.

This property resists statistical analysis attack effectively.

5.2 Information Entropy Analysis

The image information entropy reflects the distribution of image. For the information entropy of original and cipher images, we get the Lena image are 7.3897 and 7.9911, respectively. The Cameraman image are 7.0727 and 7.9912, respectively. Finally, the Elephant image are 7.3504 and 7.9912, respectively. The information entropy of these three cipher images increased obviously, which means that the distribution of cipher images are more uniform and have more random information. Note that the maximum possible information entropy for images is 8.

5.3   The Correlation Analysis of the Adjacent Pixel

We randomly select 2000 pairs of adjacent pixels in the original image and cipher image. Measure their correlation coefficient from the three directions of the horizontal, vertical and diagonal accord the below formula.

$$r_{xy} = \frac{E(xy) - E(x)E(y)}{\sqrt{D(x)}D\sqrt{(y)}} \tag{16}$$

Where E is the mathematical expectation, D is variance, the correlation of the adjacent pixel test results for the Lena and Cameraman image encryption are shown in Table 1, and the correlation effect of adjacent pixel in horizontal direction is shown in Fig. 4.

Table 1. The correlation coefficient from the three directions of the image

	Lena original image	Lena cipher image	Cameraman original image	Cameraman cipher image
Horizontal direction	0.9606	0.0052	0.9778	0.0074
Vertical direction	0.9499	0.0198	0.9793	0.0298
Diagonal direction	0.9585	0.0065	0.9686	0.0089

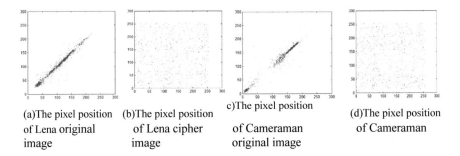

(a)The pixel position of Lena original image

(b)The pixel position of Lena cipher image

c)The pixel position of Cameraman original image

(d)The pixel position of Cameraman

Fig. 4. The correlation analysis of the adjacent pixel

From the Table 1 and Fig. 4 can see that there is a correlation between the pixels of the original image, after encrypting, the original pixel correlation is disrupted, the correlation value of the cipher image is very low.

5.4   Sensitivity Test of the Key

The sensitive of the encryption key determines the security of image encryption algorithm. In this experiment, we just change one initial iteration value $x_1(1)$ from $x_1(1) = 0.1$ to $x_1(1) = 0.099999999$, and the other five initial values where kept the same, then the three cipher images cannot be decrypted. It shows that the image encryption key is very sensitive, which can resist the enumeration attack. Based on this experiment, we can roughly estimate size of the key space. Assuming that key values are within the range −1 to 1. Also assume that a change of 0.000000001 in any of the six key values leads to a key that cannot successfully decrypt the image. Then the number possible keys is $(2/0.000000001)^6 = 6.4E55$. A powerful computer that can try 1 billion keys every second will still need 2.03E39 years to check every possible key.

5.5   Compare with the Image Encryption Algorithm Based on Logistic and Arnold Transformation

The references [13, 14] use the logistic chaotic system for image pixel value replacement and Arnold transform for pixel position shuffling. We compare our algorithm to these algorithms. Meanwhile, we also compare our algorithm to the one which uses the CNN for image pixel value replacement and Arnold transform for pixel position shuffling, the results are shown in Table 2.

Table 2.   Comparing the encryption effect of the two algorithms (IE: Information Entropy)

	Lena image		Cameraman image		Elephant image	
	NPCR	IE	NPCR	IE	NPCR	IE
Logistic-Arnold algorithm	0.9949	7.9842	0.9952	7.9863	0.9951	7.9855
CNN-Arnold algorithm	0.9944	7.9910	0.9950	7.9910	0.9955	7.9911
Our algorithm	0.9959	7.9911	0.9960	7.9912	0.9960	7.9912

From the Table 2, we can see that, comparing to the logistic-Arnold algorithm and CNN-Arnold algorithm. The security of our algorithm is better.

6   Conclusion

This image encryption algorithm is based on CNN chaotic system and W matrix transformation. Our process is firstly replace the pixel values by performing XOR operation between the original pixel values and the chaotic sequences to get the pixel value replacement image, then we change the pixel position of this image according to the W matrix which can get the final cipher image. This algorithm is simple, and the experiment results show that the NPCR value and information entropy value are higher when compared to one of the state of art existing works, this algorithm is highly useful in the field of image encryption.

Appendix (Chaos Sequence Modification Process)

Input: Original Chaos Sequences (each dimension of the 5D sequence)
Output: Modified Chaos Sequences
Parameter: N: any integer ~1000 (to make numerical value of S1 > 255)

---

1. Construct the matrix S_0, where each column is each dimension of the 5D chaotic sequence from CNN and L is the length of the sequences.

$$S_0 = \begin{bmatrix} x_1(0) & x_2(0) & \text{K} & x_5(0) \\ x_1(1) & x_2(1) & \Lambda & x_5(1) \\ \text{M} & \text{M} & \text{O} & \text{M} \\ x_1(L) & x_2(L) & \Lambda & x_5(L) \end{bmatrix}$$

2. $S_1 = \text{abs}(\text{ceil}(NS_0))$

3. Take any linear combination of the columns of S_1 to construct another matrix with 3 columns, S_2.

4. $S_3 = \text{mod}(S_2, 255)$. Each column S_3 is a modified chaotic sequence.

References

1. Dong, H.S., Lu, P., Ma, X.: Image encryption algorithm based on CNN hyper chaotic system and extend zigzag transformation. Comput. Appl. Softw. **30**(5), 133–137 (2013)
2. Sengodan, V., Balamurugan, A.: Efficient signal encryption using chaos-based system. Int. J. Electron. Eng. **2**(2), 335–338 (2010)
3. Zhong, H.Q., Li, J.M.: Image encryption scheme based on hyper chaotic sequence. Res. Comput. **30**(10), 3110–3113 (2013)
4. Telem, A.N.K., Segnig, C.M., Kenne, G., Fotsin, H.B.: A simple and robust gray image encryption scheme using chaotic logistic map and artificial neural network. Adv. Multimed. **2014**(12), 1–13 (2014)
5. Patidar, V., Pareek, N.K., Purohit, G., Sud, K.: Modified substitution diffusion image cipher using chaotic standard and logistic maps. Commun. Nonlinear Sci. Numer. Simul. **15**(10), 2755–2765 (2010)
6. Li, G.D., Zhao, G.M., Xu, W.X., Yao, S.Z.: Research on application of image encryption technology based on chaotic of cellular neural network. J. Digit. Inf. Manag. **12**(2), 151–158 (2014)
7. Chua, L.O., Yang, L.: Cellular neural networks: theory. IEEE Trans. Circuits Syst. **35**(10), 1257–1272 (1988)
8. Ren, X.X., Liao, X.F., Xiong, Y.H.: New image encryption algorithm based on cellular neural network. J. Comput. Appl. **6**(11), 1528–1535 (2011)

9. Wang, Y., Wu, C.M., Qiu, S.J.: Block encryption algorithm based on chaotic characteristics of cellular neural networks. Comput. Appl. Softw. **30**(11), 191–194 (2013)

10. Li, L.: Study on hyper chaos and hyper chaos synchronization method for cellular neural networks. Dissertation, Harbin Institute of Technology, Harbin, China (2013)

11. Wu, Y.L.: A novel transform matrix used for image scrambling. Electron. Sci. **21**(3), 69–72 (2008)

12. Kocarev, L.: Chaos-Based Cryptography, 1st edn. Springer, Berlin (2011)

13. Dureja, P., Kochhar, B.: Image encryption using Arnold's cat map and logistic map for secure transmission. Int. J. Comput. Sci. Mob. Comput. **4**(6), 194–199 (2015)

14. Singh, V., Dubey, V.: A two level image security based on Arnold transform and chaotic logistic mapping. Int. J. Adv. Res. Comput. Sci. Softw. Eng. **5**(2), 883–887 (2015)

Improved Term Weighting Factors for Keyword Extraction in Hierarchical Category Structure and Thai Text Classification

Boonthida Chiraratanasopha[1(⊠)], Thanaruk Theeramunkong[2,3], and Salin Boonbrahm[1]

[1] School of Informatics, Walailak University, Nakhon Si Thammarat, Thailand
jboontidal6@gmail.com, salil.boonbrahm@gmail.com
[2] School of ICT, Sirindhorn International Institute of Technology,
Thammasat University, Bangkok, Thailand
thanaruk@siit.tu.ac.th
[3] The Royal Society of Thailand, Bangkok, Thailand

Abstract. Keyword extraction of complex hierarchical categories becomes a challenge in text mining since commonly used classification for flat categories results in low accuracy. This paper presents a method to improve keyword extraction from hierarchical categories considering terms occurred in category from a hierarchy as additional factors in term-weighting. The method is an enhancement of a basic TF-IDF calculation; thus, it can comfortably be used for keyword extraction and classification. By taking term frequency and inverse document frequency of categories hierarchically related to a focused category, we can determine how important terms are in their family categories. In this work, hierarchy relations used in calculation are sub-categories, supercategories and sibling-categories. From experiment results, we found that the proposed method gained higher accuracy for about 40% from a baseline in a classification task.

Keywords: Keyword extraction · Term weighting · Hierarchical categories
Term frequency-inverse documents frequency (TF-IDF)

1 Introduction

Keyword extraction is a process to extract keywords with respect to their relevance in the text without prior vocabulary [1] and is an important task in the field of text mining. There are many approaches which keyword extraction can be carried out, such as supervised and unsupervised machine learning and statistical methods [2]. Most existing methods attempted a flat categories structure and applied TF-IDF [3] paradigm to weight words in order to extract important keywords [4–6] for classification, and results are commonly effective.

Unfortunately, in many circumstances, texts are organized in a form of hierarchical classes instead [7–9]. It is still new in how to extract good representative words for

© Springer Nature Switzerland AG 2019
T. Theeramunkong et al. (Eds.): iSAI-NLP 2017, AISC 807, pp. 58–67, 2019.
https://doi.org/10.1007/978-3-319-94703-7_6

texts in a hierarchical structure. The methods of current keyword extraction from a flat structure do not perform well for the tasks. Past several works [7, 10, 11] aimed to handle for classifying hierarchical categories using flat-wise extraction from each level in hierarchy structure, but the results were not impressive since they did not consider relevance among levels and relations of classes with the same root. Moreover, current methods mostly focused on classifying process for hierarchy, but they did not consider a class hierarchy affecting word frequency in extraction. Hence, they cannot yield a set of good representative keywords for categories in a hierarchy. For this reason, we will investigate in TF-IDF with relations of each category in hierarchical categories to improve the TF-IDF baseline for a hierarchical text classification task.

One of the most complex hierarchical categories is a hierarchical category set from Thai Reform website [12]. It is a website supported by Thai government agency for Thai citizen to provide opinions or suggestions on how to improve Thailand. The categories were designed to cover all aspects of problems in Thailand in hierarchy fashion and results in many categories in many dimensions. Some categories may contain overlapping concepts but in different domains. By a demand for automatic text classification of these texts, we found that it can be very difficult to find good representative terms with existing term-weighting methods such as TF-IDF. Hence, we aim to improve a keyword extraction result by considering a relation in hierarchy within a term-weighting process. The proposed method will be used for text classification for classifying hierarchical texts from Thai reform.

2 Background

2.1 Thai Reform Webpage

Thai Reform is a webpage created for receiving comments or opinions suggesting solutions for reforming Thailand. It belongs to National Reform Council of Thailand and has been supported by government agency. In the website, a citizen can post Thais textual comments freely. Moreover, the website also includes paper-based comments received from meetings, and they are transformed to machine-readable text. With all obtained comments, their experts manually assign categories fit to the opinions based on a designed hierarchical structure for a purpose of grouping and searching. The hierarchical categories structure consists of 18 top categories expanding to 3–4 level for a total of 375 leaf categories for tagging a text. For exemplifying, some parts of the hierarchical categories are translated to English and shown in Fig. 1.

From the hierarchical category structure, we can see that many categories and its sub-categories have designed closely and complexity. For example, the sub-category 6.1.1 contains a concept of *legislation, independent organizations of consumer protection* while the sub category 6.1.2 contains a concept of *laws amendment, measures development relating to consumer protection*. Moreover, the sub category 10.1.3 contains a concept of *laws and rules development of local government*. With the sharing concepts, these three categories can become very closely to one another, and very complicated. Hence, they are difficult to carelessly distinguish them even for human experts.

```
...
6. Reform Topic of Consumer Protection
          * 6.1 Organizational and personnel reform of consumer protection
                    *6.1.1 Legislation of independent organizations for consumer protection
                    *6.1.2 Law amendment or measures development with relating to consumer
                    protection
          + 6.2 System and process reform of consumer protection
...
10. Reform Topic of Local Government
          - 10.1 Organizational and personnel reform of local government
                    *10.1.1 Local government restructuring
                    *10.1.2 Decentralization to the local
                    *10.1.3 Laws and rules development of local government
          - 10.2 System and process reform of local government
                    *10.2.1 Promotion of local administration
                    *10.2.2 Encourage participation of civil society
                    *10.2.3 Improve budget allocation
```

Fig. 1. An example of Thai Reform hierarchical category structure (In this Figure, - denotes an expanded category, + denotes an expandable category with sub categories, and * denotes a leaf category.)

2.2   Related Work

In the past, there are several researches [7, 9–11, 13, 14] in hierarchical classification which can be divided into three approaches such as flat, local and global approaches. All of them tried to improve performance or developed hierarchical classification, and they are summarized as shown in Table 1.

From Table 1, all of them mentioned about using feature selection methods such as bag-of-words, n-gram features, term frequency (TF), term frequency-inverse document frequency (TF-IDF) and others. They also showed impressive results with their setup. However, we found that they did not directly use a relationship in hierarchy categories for feature selection, and most of documents were annotated with only leaf categories. Thus, the used documents were barely different from a flat category set and can comfortably be handled with a regular TF-IDF paradigm.

3   Hierarchical Term-Weighting for Keyword Extraction

This work aims to find a method to improve the accuracy of keyword extraction for Thai Reform data by considering a hierarchical relation within term-weighting calculation. We attempt to enhance existing frequent-used term weighting factors, TF-IDF, with relation of hierarchical categories on selecting keywords. The enhancement includes considering relation of terms with terms in sub categories, super categories and/or sibling categories. The overview of processes is drawn in Fig. 2.

Table 1. A summary of recent works on keywords extraction for hierarchical classification

Paper	Hierarchy feature	Feature extraction	Methods	Classification approach
[7]	Unused	– Omitted Stop word – Information Gain (IG) – Unigrams	– Flat-KNN, Flat-NB	Flat
			– Hier-ebay-struc+KNN +SVM – Latent+KNN+SVM	Local
[9]	Partial	– No word stemming – No stop-word removal – Bag-of-words, TF-IDF – Sibling Information	– PA(Flat PA)	Flat
			– Hierarchical PA (HPA) – Hierarchical PA with latent concepts (LHPA)	Global
[10]	Unused	– n-gram features – bag-of-word features – binary word features	– NB, ME, SVM	Flat
			– NB, ME, SVM – Single-path hierarchical (SPH) – Multi-path hierarchical (MPH) – Refined hierarchical classifier (RHC)	Local
[11]	Unused	– CHI-Square	– NB, SVM	Flat
			– Cosine Similarity – Exclusive Parent Training Policy (EPT) – Exclusive All Training Policy (EAT) – Exclusive Top-Level Training policy (ETT)	Flat
[13]	Unused	– TF-IDF, SVD (clustering) – TF, TF-IDF (classification) – Unigram, Bigram	– Clustering – KNN (fine-level classifier: flat) – SVM (coarse-level classifier)	Local
[14]	Partial	– TF-IDF – DeltaTF (Sibling information) – TEStDev (Standard deviation)	– K-NN – Ontology – Clustering	Flat

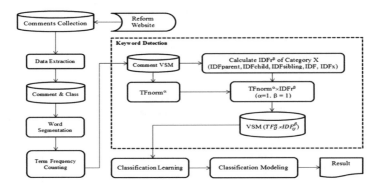

Fig. 2. A overview of Thai Keyword extraction using TFIDF with relation of hierarchical categories for Thai reform document

3.1   Pre-process

In Thai Reform webpage, comments are written in Thai texts. Some are in a short sentence while some are in lengthy details. For keyword extraction purpose, words in a comment are focused as a feature for representing a similarity of content in categories.

In this process, frequently found typos and misspelling are manually edited since they greatly affect further processes in terms of accuracy. Words in a comment are segmented using LexTo word segmentation tool [15] since it is natural for Thai texts to be inputted in consecutive manner without a space or a notable symbol between words. Non-text characters including symbols and numerical characters are removed since they represent a little to no semantic meaning in a context.

Term normalization is applied to normalize the TF weights of all terms occurring in a document by the maximum TF in the document.

3.2   Enhancing TF-IDF with Hierarchical Relations

In this process, we exploit relations of a hierarchy as a factor for term weighting. In a hierarchy, parent-child relations are considered as a super-type and sub-type respectively. Moreover, a sibling category can also be used to signify a significance of terms. Thus, we will enhance a common IDF with these relations, and we expect them to help in extracting keywords in hierarchical categories. The enhanced IDF with hierarchical relation will be mentioned for IDFr.

Let us define primary terms as follows. First, TF is a number of occurrences of each word (w) in a document (d) defined as $N(w, d)$. Basically, TF is normalized by dividing with maximum TF, $N(w, d)/Max(N(w, d))$ in word-document vector. Another feature in this work is term frequency-inverse document frequency (TF-IDF). IDF is used to represent a word that occurs in general. It will assign weight as low as possible, if it occurs in all documents [16]. These weights are then used to indicate the important keywords from documents. The equation of TF-IDF is shown in the following expression.

$$TF\text{-}IDF = N(w, d) \times \log(|D|/N(d, w)) \qquad (1)$$

When $N(w, d)$ refers to the number of occurrences of each word (w) in a document (d), while IDF is logarithmic scale value of the collection of whole document (D) divided by the number documents that appeared the word (w).

For our proposed method, we enhance a calculation of only IDF part while the TF part remains intact. There are three relationships including IDF parent, IDF child and IDF sibling relations for IDFr calculation. If category X is a top category, it has only IDF child relation. If category X is child category, it has IDF parent and IDF sibling relations. This method also applies IDF baseline in calculating for IDFr. For TF normalized-IDFr defined in Eq. (2).

$$TF \text{ normalized-IDFr} = TF \text{ normalized} \times IDF \text{ parent} \times IDF \text{ child} \times IDF \text{ sibling}$$
$$\times IDF \times IDFx \qquad (2)$$

Where IDF parent = $\log(|Dp|/N(dp, w))$, IDF parent is logarithmic scale value of the collection of parent document of category X (Dp) divided by the number parent documents of category X (dp) that appeared the word (w). The IDF child = $\log(|Dc|/N(dc, w))$, as logarithmic scale value of the collection of child document of category X (Dc) divided by the number child documents of category X (dc) that appeared the word (w). For the IDF sibling = $\log(|Ds|/N(ds, w))$, is logarithmic scale value of the collection of sibling document of category X (Ds) divided by the number sibling documents of category X (ds) that appeared the word (w) respectively. Please consider example of TF of hierarchical category structure given in Fig. 3 and TF-IDF with relations in each category X given in Table 2.

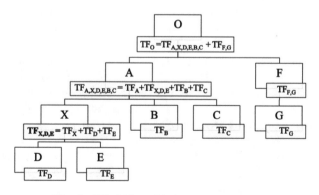

Fig. 3. TF of hierarchical category structure

From the example, the IDF weighting factors with relations in a category give more weighting factors into assign candidate terms in categories. These will obtain more related keywords in any category considering from its sub categories, super categories and/or sibling categories.

Table 2. TF and IDF with relations in each category X

No.	Weighting factors	Parent A	Sibling B, C	Children D, E	Collection O	Itself X
1	TF	$TF_{(A,X,D,E,B,C)}$	$TF_{(X,D,E,B,C)}$	$TF_{D,E}$	TF_O	$\mathbf{TF_{X,D,E}}$
2	IDF	$IDF_{(A,X,D,E,B,C)}$	$IDF_{(X,D,E,B,C)}$	$IDF_{D,E}$	IDF_O	$IDF_{X,D,E}$
3	TF-IDF	$TF_{X,D,E} \times$ $IDF_{(A,X,D,E,B,C)}$	$TF_{X,D,E} \times$ $IDF_{(X,D,E,B,C)}$	$TF_{X,D,E} \times$ $IDF_{D,E}$	$TF_{X,D,E} \times$ IDF_O	$TF_{X,D,E} \times$ $IDF_{X,D,E}$

4 Experiment

To test our proposed method, we set up an experiment to see an effect of using TF normalized-IDFr in Thai Reform classification.

4.1 Experiment Setting

In this experiment, we selected two main topics from Thai reform comment database including topic of the category#6 Reform topic of consumer protection and the category#10 Reform topic of local government for data statistics details shown in Table 3.

Table 3. Statistics of small subset of Thai reform comment database

#Doc	#Categories	#Features	Max depth
11,428	16	5,813	3

The database contains 11,428 comments with total 16 sub categories. The Thai comments were word-segmented using LexTo. For feature management, we removed non-text characters.

A comparison focuses into two aspects. The first aspect is to compare the IDFr with other term weighting factors including TF, TF normalized and TF normalizedIDF baseline. The second aspect is to apply different classification approaches to examine results. Three classification approaches i.e. Bayes classifier (NB) [17], Support Vector Machines (SVM) [18] and K-Nearest Neighbor (KNN) [19] were selected in this experiment. For setting, greedy kernel is applied for NB. A minimal bandwidth of the kernel is 0.1, and a number of kernels are set as 10. For SVM, a radial basis function is used, and a parameter of gamma and C is set as 0 and 9, respectively. Last, a parameter of K in KNN is 3, and we select numerical Manhattan distance as a measure type. We then compare the results of four weighting features and classifications based on same settings and feature amount. 5-fold cross validation is applied to test the methods. Measurements for classification are accuracy (a), precision (p) and recall (r) respectively.

4.2   Experiment Result

The results of the experiment are given in Table 4. We found that using TF normalize-IDFr with NB gain the most improved accuracy result.

Table 4. Table of the result comparing between using TF, TF normalize, TF normalized-IDF baseline and TF normalized-IDFr for classification by NB, SVM and KNN

Feature extraction	NB			SVM			KNN		
	A	P	R	A	P	R	A	P	R
TF	17.33%	17.19%	5.79%	20.97%	17.52%	12.63%	17.99%	16.30%	13.66%
TF normalized	24.40%	19.69%	19.27%	15.23%	6.51%	7.48%	17.69%	15.43%	12.79%
TF normalized-IDF baseline	20.23%	16.17%	15.40%	15.22%	14.45%	7.59%	18.51%	15.77%	13.25%
TF normalized-IDFr	63.21%	53.21%	50.99%	28.61%	18.88%	19.93%	22.44%	22.58%	16.49%

4.3   Discussion

From the results, TF normalized-IDFr weighting factor for keywords extraction on hierarchical classification shows a potential to improve performance on accuracy, precision and recall better than three traditional weighting factors. Especially classify with NB technique is the best one, when compared with SVM and KNN techniques.

The IDFr weighting factor used relation information on hierarchical categories as sub categories, super categories and/or sibling categories of any category on hierarchical categories. By a usage of IDFr, the IDF values are much more different than the IDF baseline, which includes the additional weighting factors that are IDF parent, IDF child and IDF sibling relations. Hence, the related relationships in hierarchical category structure are obvious benefits.

By observation on the incorrect classification result, we found that a major problem relied heavily on the categories from Thai reform. For example as shown in Sect. 2.1, the categories are not clearly distinctive. This issue caused a term to gain less significant in distinctive power to stand out to represent a single category in overall view. However, considering a hierarchical structure greatly helps in improving classification results. As shown in Table 4, the three common term-weighting methods apparently received unacceptably low result. The increasing in classification result with IDFr with every classification approaches shows a good sign that considering hierarchical structure of categories can assist in such difficult classification. Moreover, from the incorrect results from IDFr, we found that most of assigned categories were in family categories such as sibling categories of the same tree, unlike other term-weighting methods. Hence, the incorrect results from IDFr were not a totally off target, but slightly misses within acceptable range.

5   Conclusion

This paper proposes a method to improve term weighting features in a consideration with relationship of hierarchical categories for improving keyword extraction on hierarchical classification. An IDF is designed to enhance with information of sub-categories, super-categories and/or sibling categories, called IDFr, for signifying an inheritance of important terms representing in hierarchical structures to solve complexity in determining keywords in closely designed hierarchical categories such as data from Thai reform website. The newly invented IDFr can be used as regular term weighting along with other factors such as TF and TF-IDF for keyword extraction or classification. In this work, an experiment was set up to compare the proposed method with commonly used methods such as TF-IDF in classification task for Thai reform data. The result showed that our method the TF normalized-IDFr won over the common method in classification of hierarchical categories, and it was significantly better than the three baseline methods with NB classification approach.

Acknowledgement. Author would like to thank National Reform Council for providing comment data from Thai Reform Website.

References

1. Uzun, Y.: Keyword extraction using Naive Bayes. Department of Computer Science, Bilkent University, Turkey (2005). www.cs.bilkent.edu.tr/~guvenir/courses/CS550/Workshop/Yasin_Uzun.pdf
2. Siddiqi, S., Sharan, A.: Keyword and keyphrase extraction techniques: a literature review. Int. J. Comput. Appl. **109**, 18–23 (2015)
3. Salton, G., Buckley, C.: Term-weighting approaches in automatic text retrieval. Inf. Process. Manag. **24**, 513–523 (1988)
4. Tipsena, R.: Automatic question classification on webboard using text mining techniques. J. Sci. Technol. Mahasarakham Univ. **33**, 493 (2014). (in Thai)
5. Sarakit, P., Theeramunkong, T., Haruechaiyasak, C., Okumura, M.: Classifying emotion in Thai Youtube comments. In: 2015 6th International Conference of Information and Communication Technology for Embedded Systems (IC-ICTES), pp. 1–5. IEEE (2015)
6. Obasi, C.K., Ugwu, C.: Feature selection and vectorization in legal case documents using chi-square statistical analysis and Naïve Bayes approaches. IOSR J. Comput. Eng. **17**, 42–50 (2015)
7. Shen, D., Ruvini, J.-D., Sarwar, B.: Large-scale item categorization for e-commerce. In: Proceedings of the 21st ACM International Conference on Information and Knowledge Management, pp. 595–604. ACM (2012)
8. SillaJr, C.N., Freitas, A.A.: A survey of hierarchical classification across different application domains. Data Min. Knowl. Discov. **22**, 31–72 (2011)
9. Qiu, X., Huang, X., Liu, Z., Zhou, J.: Hierarchical text classification with latent concepts. In: Proceedings of the 49th Annual Meeting of the Association for Computational Linguistics: Human Language Technologies: Short Papers, vol. 2, pp. 598–602. Association for Computational Linguistics (2011)
10. Qu, B., Cong, G., Li, C., Sun, A., Chen, H.: An evaluation of classification models for question topic categorization. J. Am. Soc. Inf. Sci. Technol. **63**, 889–903 (2012)

11. Phachongkitphiphat, N., Vateekul, P.: An improvement of flat approach on hierarchical text classification using top-level pruning classifiers. In: 2014 11th International Joint Conference on Computer Science and Software Engineering (JCSSE), pp. 86–90. IEEE (2014)
12. Thai Reform website. http://static.thaireform.org/
13. Javed, F., Luo, Q., McNair, M., Jacob, F., Zhao, M., Carotene, K.T.: A job title classification system for the online recruitment domain. In: 2015 IEEE First International Conference on Big Data Computing Service and Applications (BigDataService), pp. 286–293. IEEE (2015)
14. Kashireddy, S.D., Gauch, S., Billah, S.M.: Automatic class labeling for CiteSeerX. In: 2013 IEEE/WIC/ACM International Joint Conferences on Web Intelligence (WI) and Intelligent Agent Technologies (IAT), pp. 241–245. IEEE (2013)
15. LexTo. http://www.sansarn.com/lexto/
16. Manning, C.D., Raghavan, P., Schütze, H.: Introduction to Information Retrieval, vol. 1. Cambridge University Press, Cambridge (2008)
17. Frank, E., Bouckaert, R.R.: Naive Bayes for text classification with unbalanced classes. In: European Conference on Principles of Data Mining and Knowledge Discovery, pp. 503–510. Springer (2006)
18. Joachims, T.: Text categorization with support vector machines: learning with many relevant features. In: Machine Learning: ECML 1998, pp. 137–142 (1998)
19. Al-Jadir, L.: Encapsulating classification in an OODBMS for data mining applications. In: Proceedings of Seventh International Conference on Database Systems for Advanced Applications, pp. 100–101. IEEE (2001)

Ontology-Based Classifiers for Wikipedia Article Quality Classification

Kanchana Saengthongpattana[1,2], Thepchai Supnithi[1],
and Nuanwan Soonthornphisaj[2(✉)]

[1] Language and Semantic Technology Laboratory,
National Electronics and Computer Technology Center (NECTEC), Pathumthani, Thailand
{kanchana.sae,thepchai.supnithi}@nectec.or.th
[2] Department of Computer Science, Faculty of Science, Kasetsart University, Bangkok, Thailand
fscinws@ku.ac.th

Abstract. Quality of Wikipedia article is the main issues that need to be solved. This research proposes the ontology-based classification framework that considers the quality of article in term of its comprehensive content which is one of the properties for featured and good articles in Thai Wikipedia. We create concepts or main ideas of articles in three domains using ontology as a knowledge representation. Knowledge based are created using OAM tool that do data mapping and classify the quality of articles via set of rules. We have investigated the ontology approach which combined Naïve Bayes classifier and found that the precision of our proposed method outperform traditional Naïve Bayes for two times.

Keywords: Ontology · Thai Wikipedia · Concept feature
Naïve Bayes Classifier

1 Introduction

Currently, the number of Thai Wikipedia articles are over 115,402 [20], whereas the number of English article reaches 5 million pages [25]. The statistics obtained from Wikipedia shows that there are only 1,139 Thai active users [28] compared to 140,999 non-Thai users [26]. These high number of articles is the main burden of active registered users to edit or classify their qualities.

Wikipedia provides editing guideline that convinces all editors to follow. This guideline can be applied to consider the level of quality. The quality assessments are mainly handled by members of WikiProjects that allow articles submission to be consider in the list of features or good article candidates. These articles will be modified by other users until they are qualified as a feature or good articles. However, this process of articles assessment cannot be fulfill by such a few number of active users. Therefore, we need a kind of automatic tool that can classify the quality of the articles.

© Springer Nature Switzerland AG 2019
T. Theeramunkong et al. (Eds.): iSAI-NLP 2017, AISC 807, pp. 68–81, 2019.
https://doi.org/10.1007/978-3-319-94703-7_7

2 Related Work

This section provides the background regarding the criteria that meet the writing standard according to Wikipedia guideline. Then we review the research works that have been done to solve the classification problems.

2.1 Standards of the Quality of Articles in Wikipedia

Wikipedia has already set up an article assessment project, which organizing the evaluation of over 115,402 Thai articles and 5,372,109 English articles into stages of quality, namely "good article", "featured article" and other quality articles. A good article is a satisfactory article that has met the good article criteria but may not have met the criteria for featured articles. A featured article [24] exemplifies the best quality of work and is distinguished by professional standards of writing, presentation, and sourcing. The description of standards are explained in Table 1.

Table 1. Standards of featured articles in Wikipedia

Criteria	Description
(1) Well-written	Its prose is engaging, even brilliant, and of a professional standard
(2) Comprehensive	It neglects no major facts or details and places the subject in context
(3) Well-researched	It is a thorough and representative survey of the relevant literature. Claims are verifiable against high-quality reliable sources and are supported by inline citations where appropriate
(4) Neutral	It presents views fairly and without bias
(5) Stable	It is not subject to ongoing edit wars and its content does not change significantly from day to day, except in response to the featured article process
(6) Appropriate- style	It follows the style guidelines, including the provision of a lead, appropriate structure and consistent citations
(7) Media	It has images that follow the image use policy and other media where appropriate, with succinct captions, and acceptable copyright status
(8) Length	It stays focused on the main topic without going into unnecessary detail

Wiki Project quality assessment initiated by Wikipedia commits to grade the levels of quality (A, GA, B, C, Start, Stub, FL and List) [27]. However, these predefined levels are not appropriate for Thai Wikipedia since there are fewer number of Thai articles and the writing contents are less comprehensive compared to those of English articles. Therefore we define two quality levels which are Feature and Good articles [21]. We focus on the comprehensive property of the content by using ontology to represent the concept or main ideas. Note that other features corresponding to criteria 6–8 are studied in [15, 16].

2.2   Wikipedia Quality Classification Research

An automatic assessment system for Wikipedia articles is useful for reader and contributors who improve the article. WikiLyzer [17] was developed as a Web-based toolkit assisting a variety of tasks about improve the quality of articles. This tool assists users to search good articles and identify the weaknesses to improve article. GreenWiki [6] evaluates the quality of articles and shows indicators that help users to get to their own conclusions about the quality of an article. Moreover, the visualizaton tools can be found in WikipediaViz [5]. It was implemented using PHP and integrated as a plugin in the Mediawiki system. This research proposed five metrics for Maturity and Quality assessment of Wikipedia articles including the visualizations to provide readers with important clues during the editing process. The result show that WikipediaViz significantly reduces the time spend for assessing the quality of articles while maintaining a good accuracy.

Several researchers studied the feature set found in the articles based on words, reference, link, *etc*. For example, [22] investigated external links and concluded that if Wikipedia is equipped with improved link control and filtering mechanisms, it will combat link rot and improve the overall quality of its external links and implicitly of its article contents. Over the past decade, most research focused on classification of the quality of Wikipedia articles by using a different feature set.

Machine learning algorithms were applied in many researches to investigate the different feature sets in order to classify the quality of Wikipedia articles. For example; textual features, meta data features, and language model features [11], length, citation, link, image [4]. Feature selection is one of the most difficult tasks because each feature has different advantages and disadvantages. [7] did not use feature on article but they applied the process of Wikipedia reviewers. They generated input vectors for the deep neural network (DNN) and achieved very high accuracy scores. The characteristics of team working of editors was studied in Spanish Wikipedia [2]. They created the attributes that describe the characteristics of the editors team who produced the articles and using decision tree algorithm. They found that the editor team's maximum efficiency and the total length of contribution are the most important indicators.

Editors plays an important role to improve the quality of articles. [13] developed several models to rank articles by using the editing relations between articles and editors. They found that using manual evaluation to assist automatic evaluation is a viable solution for the article quality assessment task on Wikipedia. [18] found that the role of editors is the contributing factor for the quality of article, since the good editors can improve the quality of articles. They proposed a method for assessing the quality of Wikipedia editors using crowdsourcing to manually detect changes of text meaning.

2.3   Classifiers and Ontology Techniques

Text classification problem that has been studied and shows promising result is Naïve Bayes Approach.

Naïve Bayes Approach. Naïve Bayes is a well-known approach since it has good performance on text categorization problem. Naïve Bayes computes conditional probabilities of the document's categories. Given a training set T containing M training instances $\{(X_1, y_1),\ldots,(X_m, y_m)\}$, where X_i is the feature vectors $\{x_{i,1},\ldots,x_{i,n}\}$ (n is the number of features), and $y_k \in \{c_1,\ldots,c_l\}$ (a set of l class).

The algorithm creates a model for the classifier, which is a hypothesis about the unknown function f for which $y = f(X)$. The classifier is later used to predict the unknown class $y_t \in \{c_1,\ldots,c_l\}$ for a new instance (X_t, y_t) with the known vector X_t.

$$y_t = \arg\max\left(P(y_k)\right)\Pi_{j=1}^{N} P\left(x_j|y_k\right) \tag{1}$$

The prior probability, $P(y_k)$, is estimated as the ratio between the number of examples belonging to class yk and the number of all examples. The conditional probability, $P(x|y)$ is the probabilities of seeing the occurrence of x_j giving class label, y_k.

For example, [9] proposed an approach to fight vandalism by extracting various features from the edits for Naïve Bayes classification. The classifier uses information regarding the editor, the sentiment of the edit, the "quality" of the edit (i.e. spelling errors), and targeted regular expressions to capture patterns common in blatant vandalism.

Ontology Approach. Ontology is a formal description of a domain of interest based on a set of individual (also called entities or objects), classes of individuals, and the relationships existing between these individuals. The logical statements on memberships of individuals in classes or relationships between individuals form a base of facts, i.e., a database. Besides, logical statements are used for specifying knowledge about the classes and relationships.

One of the most well-known ontology-based knowledge base is DBpedia [8] which represents a crowd-sourced community effort to structured information from Wikipedia and makes this information available on the Web. DBpedia provides a description of 4.0 million of things such as people, places, organizations, species, and diseases. Several research efforts that provide mechanisms for information retrieval from semantic knowledge bases have been reported [10, 14].

In addition to English Wikipedia, Ontology-based has been applied to represent knowledge for Arabic language [1]. This research builds an ontology which represents concepts and their semantic relations for a given domain. The ontological model to be applied on Arabic Wikipedia to extract for each article its semantic relations using its infobox and list of categories. The model proved its usefulness as it yielded consistent results.

For ASEAN language, Japanese Wikipedia is used as resources to build a large scale and general purpose Japanese ontology through ontology learning [19]. Wikipedia is applied as valuable resources for ontology learning focusing on is-a relationship. Besides, it's value in extracting other relationships. Moreover the hyponym can be extracted using relation in ontology.

3   Framework

In this section, we describe the ontology based classification framework that aims to classify the quality of Thai Wikipedia articles. As shown in Fig. 1, the data set (ThWiki) is preprocessed by domain separation in order to separate the articles into topics which are Biography, Place and Animal. Then the data labeling is done to get the quality label; good and low quality. After that the identification of concept using keyword matching is performed and the database is obtained. The ontology inference is operated to get the value that must be checked in the conditions of the rule sets. At this step, the quality of articles are classified based on their concept appeared in the content. We then create a vector of keywords and let naïve Bayes to reclassify the articles.

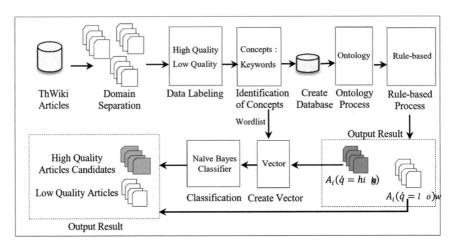

Fig. 1. The framework of the Thai Wikipedia articles quality assessment.

3.1 Data Set

Three article domains were obtained from Wikipedia (14,996 articles). The class labels are determined by considering the tags attached to each article. We consider the Featured and Good article tags as High Quality Class. The high quality means that the Wikipedia community agreed with the content quality by providing the remarkable symbol, the star, ⭐ and plus sign ⊕ to notify all users. However the low quality articles can be assured by the broom symbol, 🧹 which means that the content of article need to be rewrited or it is too short. Finally, we got the class distribution as shown in Table 2.

Table 2. Data set used in this study.

Domain	Number of articles	
	High quality	Low quality
Biography	66	8,596
Animal	9	255
Place	27	6,043
Total	102	14,894

Identification of Concepts

Our assumption is that good quality writing should have the main idea in each section of the content. Articles in Biography domain should provide information about the birth, education, occupation, family relationship, or the death concept. Articles in animal domain should provide information regarding the taxonomy, biology information. Articles in place domain should contain information related to architecture, location. Table 3 shows the list of concepts in 3 domains.

Table 3. The list of concepts in data domains.

No.	Concept	Description (Example of keywords)
		Biography domain
1	Birth	Date of birth including place of birth. ("เกิดวันที่","เกิดที่","สถานที่เกิด")
2	Knowledge	Education, training, aptitude and experties. ("จบการศึกษา","ศึกษาสาขา","เชี่ยวชาญด้าน")
3	Occupation	Occupation duty, job, and award ("ดำรงตำแหน่ง","เลื่อนยศ","ได้รับการแต่งตั้ง")
4	Relationship	Related persons, including parents, cousin, husband, wife, son, friend and partner. ("มีบุตร","อภิเษกสมรสกับ","ลูกของ","เพื่อน")
5	Death	Date of death and how he/she was death. ("สิ้นชีพ","ถูกสังหาร","อายุรวม")
		Animal domain
1	Character	Shape, elements of the body, and appearance. ("กะโหลก","สีขน","ขนาดและรูปร่าง")
2	Family/Taxonomy	About species and ethnicity. ("ข้ามสายพันธุ์","วางไข่","ออกลูกเป็นไข่")
3	Breed	The reproduction, reproduction period, including farming baby ("ต้นตระกูล","สปีชีส์")
4	Food	The food and behavior of prey. ("จับปลา","ล่าเหยื่อ","สัตว์กินเนื้อ","อาหาร")
5	Habitat	Areas where the animals can be found. ("ภายในป่า", "ถิ่นที่อยู่","อาณาเขต")
		Place domain
1	Site and Structure	Components of the place, location, set up period. ("พื้นที่ใช้สอย", "ตั้งอยู่บน","พิกัด")
2	Age	The relevant time period. Including the time that the history of that place. ("ยุคแรก", "ก่อตั้งเมือง", "ช่วงทศวรรษ")
3	Objective	The main purpose of the place. (สร้างขึ้นเพื่อ","การใช้งาน","เป้าหมาย")
4	Management	The management of the place. Organization, people involved. Methods of maintenance. Including the budget. ("ดำเนินการโดย","งบประมาณ", "องค์การบริหาร")

3.2 Create Database

We found that Wiki Code == represents the section headings and === represents subsection [23]. In our paper, "main section" is the content in the section heading. The main section consists of the subsection, paragraph as shown in Fig. 2.

== History == the inhabitants of the province before the 16th century advent of
Spanish colonisation were aboriginal peoples such as the Charrúas and the
Querandíes.
=== Contemporary history ===
In April 2013, the northeastern section of Buenos Aires Province, particularly its capital, La
Plata, experienced several flash floods that claimed the lives of at least 59 people.

Fig. 2. An example of main section.

All classes and properties in ontology are prepared as tables in a MySQL database. Four main tables are created to store data instances of Article, Subtitle, Concept, and Quality level classes. Note that the properties appeared in ontology are stored in the fields (columns) of the attached class (table). The content in main section of the article is extracted and stored in database.

Ontology
Ontology is applied to represent the concepts of each domain. Hozo [12] is used as the ontology editing tool in this work. The Ontology Web Language (OWL) which is compatible with the graphical ontology is obtained and applied during the operation with the data element. To create the knowledge base, the article in database was integrated with the ontology to create the knowledge base in the RDF (Resource Description Framework) format by using Ontology Application Management (OAM) [3] framework. RDF is a standard model for data interchange that allow data merging between different schemas. In this paper, it facilitated the ontology schema and instance of data store to be stored in OAM. In addition, the OAM semantic search application template is adopted in order to create the search system and user interface. Query process can be done on top of the provided SPARQL query facility.

To design the ontology, we must explore the data set in each domain to get the key points in term of the concepts and keywords. The relationships of these key points are determined in order to organize into the class hierarchy. Moreover the class and properties that facilitate the article retrieval process need to be included in the ontology as well. Finally the mapping between ontology and article in database make the complete knowledge base.

In this work, Classes of domain knowledge were enumerated for inclusion in the ontology which are Biography, Animal, and Place. The hierarchy of classes is determined using Is-a relation. For example, 'Position' and 'Reward' are defined as two subclasses of 'Occupation'.

There are two types of properties in this work: data properties and object properties. The data type properties describe the value type of the classes such as string, integer, boolean, *etc.* For example, the data property of the 'article name' of the 'Article biography' is string. The object properties describe the association of two related classes in the ontology. Note that it provides additional information to describe the attached class. For example, the 'Article' has two object properties: has_subtitle, and has_quality level.

The mapping between the ontology and the database to create the knowledge base can be achieved by using Ontology Application Management (OAM) software tool. The mapping configuration for each class involves three parts: class-table mapping, property-column mapping and vocabulary mapping.

In class-table mapping, user can define mapping between an ontology class and a database table.

In property-column mapping, two types of mapping can be defined: data property and object property mappings. User can define mapping between each property of a class, *i.e.* either data property or object property, with a column of a table.

The vocabulary mapping is the binding between keywords and concept class. For example, the keyword "สถานที่เกิด", "บ้านเกิด" are matched with the Class "Birth".

After the database-ontology mapping process is finished, the creation of the knowledge base in RDF format is obtained.

There are 3 ontologies in this research as shown in Figs. 3, 4 and 5.

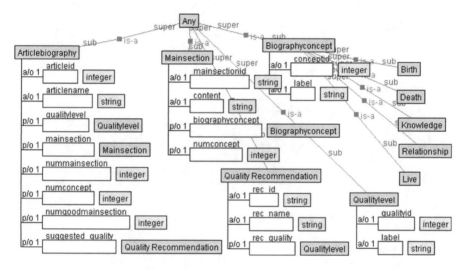

Fig. 3. Ontology of biography domain.

The structure of ontology consists of 4 main classes which are Article class, Quality level class, Main section class and Concept class. The Article class represents the description of the article in terms of article id, article name, number of concepts, quality level. Since the content of article is organized into main section so there is Main section class appeared in the ontology.

The concept class represents the main ideas in each domain. This class has a number of subclasses depending on the domain. For example, The Animal Concept consists of 4 subclasses which are character of animal, breed, family, food and habitat. The Quality recommendation contains the classified quality obtained from our approach.

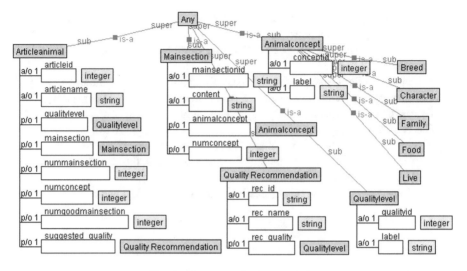

Fig. 4. Ontology of animal domain

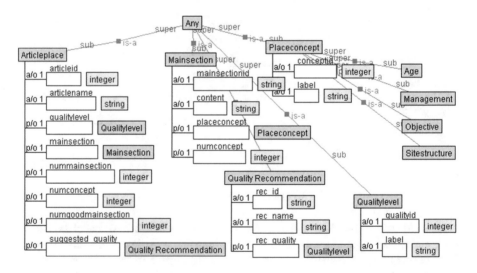

Fig. 5. Ontology of place domain

3.3 Rule-Base Process

OAM facilitates the recommender system that can be used as decision support system for user. To develop the recommender system for the Semantic Web framework, we need ontology, rules and rule-based inference engine in order to process the RDF data. The set of rules can be created using the template provided by recommendation editor. We applied the recommendation system to show the quality level results of article recommended (see Fig. 6.) in order to get the performance by comparing with the quality label.

Fig. 6. Example output of the recommendation system.

To apply the recommender system, we design the 'Recommendation Class' which contains 'attributes-of' ('rec_id', 'rec_name') and 'part-of' ('rec_quality'). Class 'Article' has 'part-of' relationship types with 'suggested_quality' which refer to 'Class Quality Recommendation'.

The rule management system is the important elements for the developing of the recommendation system. The rules used to search for articles; the different rules show different recommendation results. In this paper we study the impact of rules in fives approach as follows.

Six experimental set up are performed to explore the potential of ontology based classifier as follows:

Experiment I: We consider the content in the article as the plain text (no structure is taken into account) To determine the good quality, the article should contain at least 3 from 5 concepts in Biography and Animal domain. For those in Place domain the text should contain 2 from 4 concepts.

Experiment II: We consider the structure of the content by extracting number of concept found in each main section of the article. The 'part-of' ('numconcept', numsubtitle) of class Article and 'part-of' ('numconcept') of class 'Subtitle' are applied. The condition of rule used in this approach is the number of concepts found in each main section should be at least 1 to 3. However the total number of concepts found in the article 3.

Experiment III: The structure of article is based on each main section (the same as experiment III). However the rules are modified as we consider the total number of concepts in the article must be equal to the maximum number of domain concept (5 domain concepts for Biography and Animal domain, 4 domain concept for Place domain).

Experiment IV: We consider the ratio of keyword found in each concept in the main section based on the conditions. Figure 7 shows 3 rules. Word ratio, r_c, is the average number of word occurrence found in concept c. Given an article A, containing i main sections, the concept vector is determined by the first condition. The binary element that represent each concept is obtained by comparing r_c with the $AVG_i(r_c)$. $AVG_i(r_c)$ is the average number of word ratio r_c found in section *i*. The second rule further investigate the concept vector to level the quality of that section by considering the number of "1" found in the concept vector. If it is 2, that section is considered as a good section. After checking the second condition, we obtain the number of good sections. Then the third condition summarize the article quality by counting the number of good sections, if 60% of the total number of main sections are obtained, we get the conclusion that A is a good article. Figure 8 shows the concrete example.

The first condition:

The concept found in $A_i = \begin{cases} 1, if(r_c \geq AVG_i(r_c)). \\ 0, otherwise. \end{cases}$ As a concept vector of A_i .

The second condition:

The quality of $A_i = \begin{cases} good, if(the\ sum\ of\ the\ concept\ vector\ of\ A_i\ \leq 2). \\ not\ good, otherwise. \end{cases}$

The third condition: $(n\ is\ the\ total\ number\ of\ A_i)$

The quality of *artice A*= $\begin{cases} low, if(n\ \leq 2). \\ high, if(n > 2) \wedge (the\ number\ of\ \ A_i\ as\ the\ good\ section \geq 60\%\ of\ n) \\ low, otherwise. \end{cases}$

Fig. 7. The condition for Experiment IV and Example.

Article A		The num of word occurrence found in concept c				
n=3	#words	c1	c2	c3	c4	c5
main secti=1	**993**	0	**4**	1	8	10
main secti=2	1252	0	5	7	0	3
main secti=3	509	0	5	2	1	2

Article A	rc					AVG$_i$ (r_c)
n=3	c1	c2	c3	c4	c5	
main secti=1	0	**0.004**	0.001	0.008	0.010	0.005
main secti=2	0	0.004	0.006	0	0.002	0.002
main secti=3	0	0.010	0.004	0.002	0.004	0.004

$\frac{4}{993}$ $\quad \frac{0 + 0.004 + 0.001 + 0.008 + 0.010}{5}$

The first condition

0.004 < 0.005

Article *A*	The concept found in A_i					The quality of A_i ; sum (Aic) ≤ 2
n=3	c_1	c2	c3	c4	c5	
main sect$_{i=1}$	0	0	0	1	1	good
main sect$_{i=2}$	0	1	1	0	1	not good
main sect$_{i=3}$	0	1	1	0	1	not good

The second condition
0+0+0+1+1 = 2

The third condition,
n=3, n > 2, num sect good = 1
Article A is low quality.

Fig. 8. The example of the condition for Experiment IV

Experiment V (Baseline algorithm): We use Naïve Bayes as a baseline algorithm.

Experiment VI: This experiment aim to investigate the performance of our proposed ontology framework that use two step of quality classification. The first step classification is the contribution of set of rules. The second step is the reclassification that is done by Naïve Bayes.

4 Experimental Result

To evaluate the performance of our approach, we use Precision and F-Measure as shown in Eqs. 2 and 3. Ten-fold cross validation is performed to validate the performance.

$$Precision = \frac{No.\ of\ True\ Positive}{No.\ of\ True\ Positive\ +\ No.\ of\ False\ Positive} \tag{2}$$

$$F - Measure = \frac{2xPrecisionxRecall}{Precision + Recall} \tag{3}$$

Bernoulli Naïve Bayes is applied for second classification step. The feature vectors created for naïve bayes is $\{x_{i,1},...,x_{i,n}\}$ where n is the number of features based on keywords of concept. The value in this vector is 1 if the keyword occurs in articles and 0 otherwise.

All experimental results measured in terms of precision and F-measure are shown in Fig. 9. We found that our proposed method outperforms Naïve Bayes and gets the highest performance. The precision of ontology based classifier is 0.239 in Biography domain, 0.75 in Animal domain, and 0.174 in Place domain. For Biography domain and Place domain, the performance of ontology based classifier is two time higher than Naïve Bayes Approach. Performance of the rule-based experiment (I, II, III, IV) based on OAM tool that recommend the high quality of articles is less than Naïve Bayes Approach except for Approach III in Biography domain with a precision of 0.128 (NB is 0.105). Experiment I provides the lowest performance, therefore, it is confirmed that the structure of articles is an important factor to determine the quality.

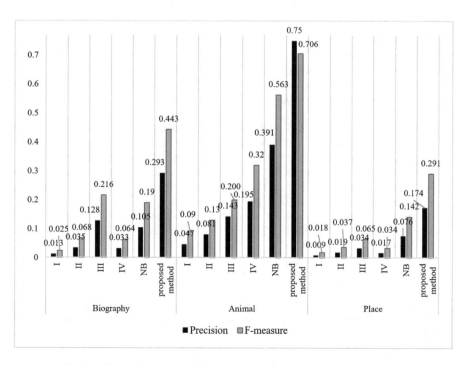

Fig. 9. The performance of all approaches in the high quality of articles.

5   Conclusion

We have investigated the ontology of Thai Wikipedia and used OAM tools for the rulesbased approach. In addition, we combined the rules-based approach with Naïve Bayes algorithms. Our proposed method obtain the highest performance. However, the rulesbased and the list of keywords in each concept may not cover all features of good content for high quality articles. For future work, we will improve the ontology and concept to produce the list of feature or good article candidates. Moreover, we will integrate an automatic concept extraction technique for other domains.

Acknowledgements. This research is supported by Kasetsart University Research and Development. (KURDI). We would like to thank Language and Semantic Technology Laboratory, NECTEC for providing the platform for ontology development. Special thank goes to Dr. Marut Buranarach for his technical support.

References

1. Al-Rajebah, N.I., Al-Khalifa, H.S., Al-Salman, A.M.S.: Exploiting Arabic Wikipedia for automatic ontology generation: a proposed approach. In: International Conference on Semantic Technology and Information Retrieval (STAIR), pp. 70–76 (2011)
2. Betancourt, G.G., Segnine, A., Trabuco, C., Rezgui, A., Jullien, N.: Mining team characteristics to predict Wikipedia article quality. In: Proceedings of the 12th International Symposium on Open Collaboration, pp. 1–9. ACM, Berlin (2016)
3. Buranarach, M., Supnithi, T., Thein, Y.M., Ruangrajitpakorn, T., Rattanasawad, T., Wongpatikaseree, K., Lim, A.O., Tan, Y., Assawamakin, A.: OAM: an ontology application management framework for simplifying ontology-based semantic web application development. Int. J. Softw. Eng. Knowl. Eng. **26**, 115–146 (2016)
4. Calzada, G.D.l., Dekhtyar, A.: On measuring the quality of Wikipedia articles. In: Proceedings of the 4th Workshop on Information Credibility, pp. 11–18. ACM, Raleigh (2010)
5. Chevalier, F., Huot, S., Fekete, J.D.: WikipediaViz: conveying article quality for casual Wikipedia readers. In: IEEE Pacific Visualization Symposium, pp. 49–56 (2010)
6. Dalip, D.H., Santos, R.L., Renn, D., Oliveira, Val, Amaral, r.F., Andr, M., Gon, alves, Prates, R.O., Minardi, R.C.M., Almeida, J.M.d.: GreenWiki: a tool to support users' assessment of the quality of Wikipedia articles. In: 11th Annual International ACM/IEEE Joint Conference on Digital Libraries, pp. 469–470. ACM, Ottawa (2011)
7. Dang, Q.V., Ignat, C.-L.: Quality assessment of Wikipedia articles without feature engineering. In: 16th ACM/IEEE-CS on Joint Conference on Digital Libraries, pp. 27–30. ACM, Newark (2016)
8. DBpedia. http://wiki.dbpedia.org
9. Harpalani, M., Phumprao, T., Bassi, M., Hart, M., Johnson, R..: Wiki Vandalysis - Wikipedia vandalism analysis. In: Lab Report for PAN at CLEF (2010)
10. Hartig, O.: SQUIN: a traversal based query execution system for the web of linked data. In: ACM SIGMOD International Conference on Management of Data, pp. 1081–1084. ACM, New York (2013)

11. Javanmardi, S.: Measuring Content Quality in User Generated Content Systems: A Machine Learning Approach. Information and Computer Science University of California, Irvine, ProQuest Dissertations (2011)
12. Kozaki, K., Kitamura, Y., Ikeda, M., Mizoguchi, R.: Hozo: an environment for building/using ontologies based on a fundamental consideration of "Role" and "Relationship". In: 13th International Conference of Knowledge Engineering and Knowledge Management: Ontologies and the Semantic Web, pp. 213–218. Springer, Heidelberg (2002)
13. Li, X., Tang, J., Wang, T., Luo, Z., de Rijke, M.: Automatically assessing Wikipedia article quality by exploiting article–editor networks. In: 37th European Conference on IR Research, pp. 574–580, Springer, Cham (2015)
14. Pan, J.Z., Thomas, E., Sleeman, D.: Ontosearch2: searching and querying web ontologies. In: Proceedings of the IADIS International Conference, pp. 211–218 (2006)
15. Saengthongpattana, K., Soonthornphisaj, N.: Thai Wikipedia quality measurement using fuzzy logic. In: The 26th Annual Conference of the Japanese Society for Artificial Intelligence (2012)
16. Saengthongpattana, K., Soonthornphisaj, N.: Assessing the quality of Thai Wikipedia articles using concept and statistical features. In: Rocha, Á., Correia, A.M., Tan. F.B., Stroetmann, K.A. (eds.) New Perspectives in Information Systems and Technologies, vol. 1, pp. 513–523. Springer, Cham (2014)
17. Sciascio, C.d., Strohmaier, D., Errecalde, M., Veas, E.: WikiLyzer: interactive information quality assessment in Wikipedia. In: Proceedings of the 22nd International Conference on Intelligent User Interfaces, pp. 377–388. ACM, Limassol (2017)
18. Suzuki, Y., Nakamura, S.: Assessing the quality of Wikipedia editors through crowdsourcing. In: Proceedings of the 25th International Conference Companion on World Wide Web, pp. 1001–1006. Canada (2016)
19. Tamagawa, S., Sakurai, S., Tejima, T., Morita, T., Izumi, N., Yamaguchi, T.: Learning a large scale of ontology from Japanese Wikipedia. In: International Conference on Web Intelligence and Intelligent Agent Technology. pp. 279–286. IEEE/WIC/ACM (2010)
20. หน้าหลัก. https://th.wikipedia.org/wiki/หน้าหลัก
21. วิกิพีเดีย:บทความคัดสรร. https://th.wikipedia.org/wiki/วิกิพีเดีย:บทความคัดสรร
22. Tzekou, P., Stamou, S., Kirtsis, N., Zotos, N.: Quality assessment of Wikipedia external links. In: The 7th International Conference on Web Information Systems and Technologies, pp. 248–254. Noordwijkerhout, Netherlands (2011)
23. Help: Wikitext Examples. https://meta.wikimedia.org/wiki/Help:Wikitext_examples
24. Featured_article_criteria. https://en.wikipedia.org/wiki/Wikipedia:Featured_article_criteria
25. Main_Page. https://en.wikipedia.org/wiki/Main_Page
26. Statistics. https://en.wikipedia.org/wiki/Special:Statistics
27. WikiProjectassessment. https://en.wikipedia.org/wiki/Wikipedia:WikiProject_assessment
28. สถิติ. https://th.wikipedia.org/wiki/พิเศษ:สถิติ

Thai Named-Entity Recognition Using Variational Long Short-Term Memory with Conditional Random Field

Can Udomcharoenchaikit[1]([⊠]), Peerapon Vateekul[1],
and Prachya Boonkwan[2]

[1] Faculty of Engineering, Department of Computer Engineering,
Chulalongkorn University, Bangkok, Thailand
can.u@student.chula.ac.th, peerapon.v@chula.ac.th
[2] NECTEC, Language and Semantic Technology Lab (LST),
Pathumthani, Thailand
prachya.boonkwan@nectec.or.th

Abstract. Thai Named-Entity Recognition (NER) is a difficult task due to the characteristics of Thai language such as the lack of special character that separates named-entity from other word types. Previous Thai NER system heavily depends on human's knowledge in a form of feature selection and external resources such as dictionaries. A recent trend in NER research shows that deep learning approach can be used to train high-quality NER system without resorting on these external resources. In this paper, we present a deep learning model that combines recurrent neural networks with a probabilistic graphical model, as well as, a variational inference-based dropout approach. We benchmark our model on one of the largest Thai corpora called "BEST 2010". Our model outperforms all baseline methods without relying on extra manually annotated resources and external knowledge.

Keywords: Named-Entity Recognition · Natural Language Processing
Deep learning

1 Introduction

Named-Entity Recognition (NER), a task of locating and identifying named entities into pre-defined categories, is essential to many Natural Language Processing (NLP) applications. Many NER systems rely on manually annotated resources such as gazetteers, etc. These resources are expensive and laborious to create. They require large amount of work from linguists and specialists from a specific domain. For Thai language, the NER task is more challenging due to the characteristics of Thai language such as the lack of orthogonal information, the lack of delimiter between words, etc. Therefore, Thai NER is more dependent on external resources provided by linguists and specialists. A recent trend in NER research shows that deep learning techniques can be used to train a robust NER model without resorting on manually annotated resources.

© Springer Nature Switzerland AG 2019
T. Theeramunkong et al. (Eds.): iSAI-NLP 2017, AISC 807, pp. 82–92, 2019.
https://doi.org/10.1007/978-3-319-94703-7_8

In Thai NER research, the existing NER systems rely heavily on specialized resources to perform well. Previous research on Thai NER mainly focuses on feature selection; many features used in these NER systems (e.g. syllable lists, keyword lists, etc.) are curated by linguists. In addition, deep learning techniques have not been implemented for Thai NER task.

In this paper, we present a deep learning model for Thai NER task—Variational Bidirectional Long Short-Term Memory with Conditional Random Field (V-BLSTM-CRF). Our model learns from sequential input data using Bidirectional Long Short-Term Memory (BLSTM) architecture, and it also learns to infer information from multiple interdependent output labels using CRFs method. A variational inference-based dropout technique is used to regularize our model. Our method has shown to be robust and perform well without resorting on manually annotated resources.

The remainder of this paper is organized as follows: the following section discusses related literature and background knowledge; Sect. 3 presents our V-BLSTM-CRF model for Thai NER task; Sect. 4 discusses the dataset used in this experiment; Sect. 5 examines the experimental results; and Sect. 6 discusses the conclusion of this research.

2   Background

2.1   Thai Named-Entities

Many challenges in Thai NER come from the characteristics of Thai language. Thai language has no special character that separates NE from other word type. For example, in English, the first character of a NE is capitalized. The lack of delimiter between words in Thai language makes it harder to define word boundary. In addition, Thai NE can be very long and can be composed of many morphemes. e.g. " คณะกรรมการกิจการกระจายเสียง กิจการโทรทัศน์ และกิจการโทรคมนาคมแห่งชาติ" (National Broadcasting and Telecommunications Commission). Despite many challenges, Thai language also contains clue words for NEs such as "องค์กร" (organization), "สำนักงาน" (office), "นาย" (mister), etc.

2.2   Statistical Sequence Labeling

During 2000's, there were multiple works focusing on Thai named-entity recognition using machine learning approach such as Maximum Entropy(ME) [1], Support Vector Machine(SVM) [2], Conditional Random Field (CRF) [3, 4], etc. They rely on specialized knowledge resources to perform NER task. None of these research papers adopt a neural network approach. The advancement in deep learning research in 2010's leads to breakthrough in many areas, NLP included. Many NLP tasks, including NER, can be done with a better performance using deep learning approach [5–8].The following sections discuss techniques used in this research.

Conditional Random Fields. Conditional Random Fields (CRFs) are a framework for creating probabilistic graphical models for annotating and segmenting sequential data [9]. CRFs learn the context from neighboring samples to form a predictive model. In previous work on Thai NER, the most recent technique is based on CRFs [3, 4]. The strength of probabilistic graphical models, including CRFs, lies in their ability to infer information from multiple interdependent variables. In linear-chain CRFs, the outputs are linked together to form a linear chain. Linear-chain CRFs, for NER task, are composed of two main factors: one factor represents the dependency between output labels, and another factor that represents the dependency between an output and its input features. The probability distribution, which can be represented by CRFs, has the form [10]:

$$P(Y|X) = \frac{1}{Z(X)} \tilde{P}(Y, X) \tag{1}$$

where $\tilde{P}(Y, X) = \prod_{i=1}^{k-1} \phi(Y_i, X_i)$ and $Z(X) = \sum_Y \tilde{P}(Y, X)$

Further details are discussed in [9, 10]. Since one of the recent works in Thai Named-Entity Recognition is based on CRFs, the baseline algorithm for this paper is, therefore, based on CRFs.

Long Short-Term Memory (LSTM). LSTM is one of Recurrent Neural Network (RNN) architectures. RNN is a type of neural network that is designed for processing sequential information. However, as the temporal span between dependencies grows, traditional RNNs become increasingly inefficient in term of representing such dependencies. This problem is also known as the problem of long-term dependencies [11]. LSTM architecture is designed to overcome this issue by incorporating a new structure called memory cell. LSTM adds or removes information to or from the memory cell by using several gates to control changes to memory cell. Therefore, LSTM is able to allow the constant error to flow through memory cells bridging long-term dependencies together [12]. Recent research in NER has shown that LSTM can be the main component for a robust NER system [7, 8].

LSTM-CRF LSTM can learn to represent dependencies across the inputs really well, but a usual softmax output layer does not take dependencies across output labels into consideration. Hence, we can incorporate CRFs into our model as an output layer to include dependencies output labels into the model. In NER task, named-entity tags have a strong dependency with each other (e.g. I-PER can only follow B-PER). Currently, LSTM-CRF model for NER [8] has shown to obtain the state-of-the-art performance across 4 languages (English, German, Dutch, and Spanish) without relying on external knowledge and resources such as gazetteers.

Word Embedding. Word Embedding is a group of natural language processing techniques, which is capable of learning high-quality word representations in a geometric space. Techniques such as Continuous Bag-of-Words and Skip-Gram have shown to provide the state-of-the-art performance on representing syntactic and semantic meaning [13]. In recent NER research, pre-trained word embedding is used [7, 8]. In this paper word embedding technique mentioned in [14] is used to learn word representations from scratch.

Regularization. Overfitting is a serious issue in deep neural networks. Dropout is a common way of preventing deep neural networks from overfitting. Dropout addresses this problem by randomly drop neurons and their connections from the network during training [15]. However, the technique does not work as well on Recurrent Neural Networks. A new dropout variant based on Bayesian interpretation, also known as Variational Inference-based dropout, has shown that instead of putting dropout layer at the inputs or the outputs of RNNs, dropout can be applied in RNNs by randomly dropping the same network units at each time step for inputs, outputs, and recurrent layers [14]. This regularization method has shown to improve performance for RNN based models. In Fig. 1, each circle represents a LSTM unit. Dashed arrows represent connections without dropout. Colored arrows represent connections with dropout, where each color represent different dropout mask. Verticle arrows represent input and output to each LSTM unit, while horizontal arrows represent recurrent connection between each LSTM unit. Variational inference-based dropout uses an identical dropout mask at each time-step; while traditional dropout uses a different dropout mask. Variational inference-based dropout also applies dropout mask at the recurrent layer.

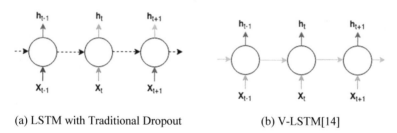

(a) LSTM with Traditional Dropout   (b) V-LSTM[14]

Fig. 1. (a) LSTM with traditional dropout technique and (b) LSTM with varational inference-based dropout technique

3   Variational Bidirectional LSTM-CRF

In this paper, we propose a new deep neural networks model for NER task. It is based on an existing LSTM-CRF model [8]. We also implement a regularization technique shown in [14] to improve the performance; "variational LSTM" refers to LSTM with regularization technique mentioned in [14]. Our method has shown to outperform CRFs, LSTM, and LSTM-CRF models on Thai NER task. Each input text sequence is padded with zeros if there are less than 2,000 words in the input sequence, or separated into smaller input sequences if there are more than 2,000 words in the input sequence. Hence, each input sequence is 2,000 words long (including padding). These discrete input features are then transformed into continuous vector representations. As shown in Fig. 1, the model uses Word Embedding technique to embedded both words (128 dimensions) and POS tags (32 dimensions), then concatenate and use them as input features for LSTM. Word embedding process can be seen as the product of one-hot encoded input vector with an embedding matrix $W_E \in R^{V \times D}$ — where D represents the size of embedding dimension and V represents the size of the vocabulary—which we yield a word vector.

LSTM is a deep learning model that is capable of learning long sequential data. Bidirectional architecture is chosen to increase the amount of input information for the output layer. When two LSTM layers of opposite direction are connected together, the output layer can adopt information from both previous and future time-steps. The forward LSTM and the backward LSTM are merged together by concatenation, then the outputs are passed to the CRF output layer. Our experiment also confirms that bidirectional architecture is able to yield a better performance.

Variational inference-based dropout is used as a regularization method in this model. In each time step, the same network units are dropped. Variational LSTM model can be formulated as [14]:

$$\begin{pmatrix} \mathbf{i} \\ \mathbf{f} \\ \mathbf{o} \\ \mathbf{g} \end{pmatrix} = \begin{pmatrix} \text{sigm} \\ \text{sigm} \\ \text{sigm} \\ \text{sigm} \end{pmatrix} \left(\begin{pmatrix} \mathbf{x}_t \circ \mathbf{z}_x \\ \mathbf{h}_{t-1} \circ \mathbf{z}_h \end{pmatrix} \cdot \mathbf{W} \right) \tag{2}$$

where i, f, o, and g represent input, forget, output and input modulation gates respectively. "sigm" refers to sigmoid. Where $\mathbf{W} = (\mathbf{W}_i, \mathbf{U}_i, \mathbf{W}_f, \mathbf{U}_f, \mathbf{W}_o, \mathbf{U}_o, \mathbf{W}_g, \mathbf{U}_g)$ weight matrices. z_x and z_h refer to dropout apply to x_t and h_{t-1} by the element-wise product (Fig. 2).

Our experiment shows that variational inference-based dropout [14] is superior to normal dropout method [15].

In Sect. 5, this model is tested against other models such as CRFs, Unidirectional LSTM (forward), Bidirectional LSTM, and BLSTM-CRF (Table 1).

4 Dataset

Benchmark for Enhancing the Standard of Thai language processing, as known as BEST2010, was a competition held by National Electronics and Computer Technology Center (NECTEC) in 2010 to find the best Thai word segmentation algorithm. NECTEC also released a dataset for the competitors which contains part-of-speech tags and location of each named-entity. However, the full version of this dataset also contains a pre-defined category of each named entity which can be used for NamedEntity Recognition task. The full version of BEST 2010 dataset can be obtained from NECTEC for research purposes. BEST2010 training set contains 5,238 text files which compose of 2,924,433 words. The test set contains 249 text files and 227,302 words. There are 37 named-entity classes in BEST 2010 dataset including 'O' which represents a non-named-entity[1]. Named-entity types in this dataset extend beyond person, organization, and location; BEST2010 also includes other named-entity classes such as

---

[1] Note that each entity is tagged with BIO format. Therefore, each named-entity can be separated into two classes. For example, B-PER and I-PER are two different classes.

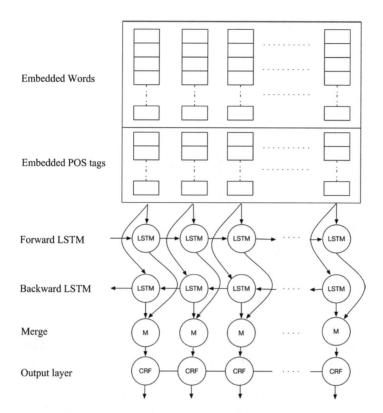

Fig. 2. Variational LSTM-CRF model for Thai Named-Entity Recognition

Table 1. List of NER models

Abbreviation	Description
CRF	Conditional Random Fields
LSTM	Unidirectional Long Short-Term Memory
BLSTM	Bidirectional LSTM
BLSTM-CRF	Bidirectional LSTM + CRF
V-BLSTM-CRF	Variational BLSTM-CRF

abbreviation, title, designation, date and time, brand, measurement, and terminology as shown in Table 2. BEST 2010 dataset tags each named-entity based on the guideline shown in [16]. There are several words in the dataset which are not tagged with any named-entity category mentioned in [16], such as NAME B and _ [2].

---

[2] These NE tags are not mentioned in [16], and BEST2010 documentation does not provide further explanation.

Table 2. Description of named-entity categories

Tag	Description
TTL	Title – a word that indicates social relationship or a permanent title
DES	Designation – position or job title
PER	Person
ORG	Organization
LOC	Location
DTM	Date and Time
BRN	Brand – Names of products or trademarks
MEA	Measurement – quantity of things in a standard unit
NUM	Number – quantity which cannot be labelled as MEA
TRM	Terminology
ABB X	Abbreviation of X named entity category
O	A word which is not a named-entity

5   Experimental Results

This paper explores a deep learning approach to Thai NER, as well as, the effect of manually annotated resource on Thai NER task. Various variations of LSTM architectures were used in this research. The robustness of each model, when POS tags are not provided, is also tested. The proposed model is the result of the following experiments. The following sections discuss the effect of various neural network variations, as well as, their performances against the existing method (CRF). Macro f-score is our main evaluation metric. The results of this paper are shown in Table 3.

Table 3. Macro precision (P), recall (R), f-score (Ma-F1), and micro f-score (Mi-F1) of NER models. The highest score of each metric is bolded.

Model	Without POS				With POS			
	P	R	Ma-F1	Mi-F1	P	R	Ma-F1	Mi-F1
CRF (baseline)	58.8	51.4	54.9	70.7	64.9	58.5	61.5	78.5
LSTM	63.8	53.5	58.2	73.8	63.4	55.2	59.0	75.3
BLSTM	65.2	59.2	62.1	79.0	67.6	61.0	64.1	80.9
BLSTM-CRF	68.5	57.8	62.7	80.9	69.9	62.0	65.7	83.6
V-BLSTM-CRF	69.6	59.1	63.9	80.7	72.8	63.5	67.8	83.7

5.1   Experiment 1: The Effect of Bidirectional Architecture

This experiment reveals that bidirectional LSTM architecture outperforms unidirectional LSTM (forward) architecture in both scenarios, with POS tags and without POS tags as input features. With POS tags, bidirectional LSTM yields a macro F1-score of 64.3 comparing to 59.01 in unidirectional LSTM. Without POS tags, bidirectional LSTM yields a macro F1-score of 62.2; while unidirectional LSTM yields a macro F1-

score of 58.2. This experiment shows that bidirectional architecture can use contextual information from previous and future time-steps to generate a better prediction.

5.2   Experiment 2: The Effect of CRFs Output Layer

Experiment 1 shows that bidirectional LSTM performs better than unidirectional LSTM. In this experiment, each model is built based on BLSTM model. In addition, we replace softmax output layer with CRFs output layer to test BLSTM-CRF architecture on Thai NER task. This experiment shows that BLSTM-CRF architecture gives a superior performance comparing to BLSTM architecture with and without POS tags as input features. Macro F1-score increases from 62.1 to 62.7 for BLSM-CRF without POS tags as input features, and from 64.3 to 65.7 for BLSM-CRF with POS tags as input features. This experiment shows that CRFs output layer can apply dependencies across output labels to make a better prediction.

5.3   Experiment 3: The Effect of Variational Inference-Based Dropout Method

Experiment 2 concludes that BLSTM-CRF model is superior to BLSTM in Thai NER task. In this experiment, we extend BLSTM-CRF model by using variational inferencebased dropout method. This experiment shows that V-BLSTM-CRF model outperforms BLSTM-CRF model in both scenarios with and without POS tags as input feature. The macro f1-score increases when using variational inference-based dropout instead of traditional dropout from 65.7 to 67.8, and from 62.7 to 63.9 for the system with POS tags and the system without POS tags respectively. Experiment 3 concludes that variational inference-based dropout [14] outperforms the traditional dropout technique.

5.4   Experiment 4: Comparison with the Existing Method

Experiment 4 compares deep learning methods proposed in this research with the baseline system based on existing research in Thai NER.

In previous work on Thai NER, the most recent technique is based on Conditional Random Fields algorithm [3, 4]. However, most of its features are derived from the "dictionaries" and "keyword lists". Since these features are not publicly available, we cannot use them as our features. In this experiment, we implemented a baseline system based on Conditional Random Fields (CRF) algorithm, which performs well in previous research on Thai NER. Our baseline CRF algorithm uses the following features: previous word, previous POS tag, next word, current word, last 2 characters of the current word, last 3 character of the current word, and is current word a number? (True/False).

The results, as shown in Table 3, reveal that V-BLSTM-CRF has the best performance. Comparing to the baseline model, unidirectional LSTM is the weakest model and it is the only deep learning model that performs worse than the baseline model when the POS tags are used. While POS tags improve the overall performance, its improvement is marginal for deep learning models. Manual POS tagging is an expensive and labor-intensive task. With an absence of POS tags, a deep learning approach is robust enough for Named-Entity Recognition task, while the performance of the model with CRFs drop drastically from 61.5 to 54.9 for macro F1-score and from 78.5 to 70.7 for micro F1-score. Hence, models with deep learning approach can learn from words and their context, and can yield more representational power from just textual data alone. In addition, POS tagging can also be done with computer at a satisfactory level.

As shown in Table 3, micro f-score for each model is much higher than its macro f-score. This indicates that our models are biased toward classes with more samples. Note that there are 31 classes in total when we ignore categories that do not exist in the test set as well as category 'O' which refers to a word that is not a named-entity. In some classes, the number of samples is lower than ten. While the largest namedentity class has 71,122 samples in our training set. Our models do not perform well on named-entity classes that have very few samples in the training set such as, NAME B, ABB TTL I, ABB LOC I and. The number of training samples of each class can be found in Table 4. The correlation score between the number of training samples and the f1-score of V-BLSTM-CRF model (with POS) is 0.4767, therefore there is a positive relationship between the number of training samples and the f1-score. In order to improve the performance of named-entity classes which have very few training samples, we need to extend the corpus to include more samples.

Table 4. List of named-entity categories in the training set sorted by their quantity in descending order and their f1-score in the test set (V-BLSTM-CRF model).

NE	# training	F1	NE	# training	F!
O	2,436,458	N/A[a]	ABB B	2,403	0.61
ORG I	71,122	0.82	ABB TTL B	1,842	0.96
PER B	59,411	0.94	TRM B	1,259	0.91
ORG B	51,494	0.78	ABB ORG I	716	0.19
MEA B	44,144	0.67	TRM I	700	0.94
DTM I	43,098	0.94	BRN B	571	0.39
LOC B	37,737	0.80	TTL I	480	0.25
LOC I	29,842	0.84	ABB DES I	472	0.69
PER I	27,015	0.97	BRN I	150	0.20
TTL B	25,238	0.98	ABB LOC I	126	0.00
DTM B	20,500	0.87	NAME B	14	0.00
MEA I	19,237	0.63	ABB TTL I	8	0.00
NUM B	18,471	0.63	–	5	0.00
ABB DES B	14,939	0.95	ABB I	3	N/A[b]
ABB ORG B	13,323	0.88	ABB PER B 2	2	N/A[†]
DES B	8,429	0.70	DDEM	2	N/A[†]
ABB LOC B	7,517	0.93	ABB	1	N/A[†]
DES I	5,495	0.66	ABB PER I	1	N/A[†]
NUM I	2,641	0.78			

[a]Note that the category 'O' refers to a word that is not a named entity, we do not include this category in our evaluation.
[b]Note that, when evaluating, we ignore categories that do not exist in the test set.

6   Conclusion

This paper presents a deep learning architecture for NER task that can perform well even without POS tags and other external resources that are expensive to obtain. Using POS as a feature contributes to the improvement of performance in NER task; however, deep learning models in this experiment can give a satisfactory result without it. Three components in our deep learning model that enhance its predictive power are: bidirectional architecture, CRFs output layer, and variational inference-based dropout. Bidirectional architecture allows contextual information from preceding and following words to be used; CRFs output layer uses dependencies across output labels to make prediction; variational inference-based dropout prevents the model from overfitting. In the future, we will explore unsupervised learning techniques to extract extra features that can improve the performance of the proposed NER model, and therefore creating a NER system which is less dependent on hand-crafted features and linguistic knowledge.

References

1. Chanlekha, H., Kawtrakul, A.: Thai named entity extraction by incorporating maximum entropy model with simple heuristic information. In: Proceedings of the IJCNLP (2004)
2. Suwanno, N., Suzuki, Y., Yamazaki, H.: Selecting the optimal feature sets for Thai named entity extraction. In: Proceedings of ICEE-2007 & PEC, vol. 5 (2007)
3. Tirasaroj, N., Aroonmanakun, W.: Thai named entity recognition based on conditional random fields. In: Eighth International Symposium on Natural Language Processing, SNLP 2009, pp. 216–220. IEEE (2009)
4. Tirasaroj, N., Aroonmanakun, W.: The effect of answer patterns for supervised named entity recognition in Thai. In: PACLIC 2011, pp. 392–399 (2011)
5. Collobert, J., Weston, L., Bottou, M., Karlen, M., Kavukcuoglu, K., Kuksa, P.: Natural language processing (almost) from scratch. J. Mach. Learn. Res. **12**, 2493–2537 (2011)
6. Huang, Z., Xu, W., Yu, K.: Bidirectional LSTM-CRF models for sequence tagging. arXiv preprint arXiv:1508.01991 (2015)
7. Chiu, J.P., Nichols, E.: Named entity recognition with bidirectional LSTM-CNNs. Trans. Assoc. Comput. Linguist. **4**, 357–370 (2016)
8. Lample, G., Ballesteros, M., Subramanian, S., Kawakami, K., Dyer, C.: Neural architectures for named entity recognition. arXiv preprint arXiv:1603.01360 (2016)
9. Lafferty, J., et al.: Conditional random fields: probabilistic models for segmenting and labeling sequence data. In: Proceedings of the Eighteenth International Conference on Machine Learning, ICML, vol. 1, pp. 282–289 (2001)
10. Koller, D., Friedman, N.: Probabilistic Graphical Models: Principles and Techniques. MIT press, Cambridge (2009)
11. Bengio, Y., Simard, P., Frasconi, P.: Learning long-term dependencies with gradient descent is difficult. IEEE Trans. Neural Netw. **5**(2), 157–166 (1994)
12. Hochreiter, S., Schmidhuber, J.: Long short-term memory. Neural Comput. **9**(8), 1735–1780 (1997)
13. Mikolov, T., Chen, K., Corrado, G., Dean, J.: Efficient estimation of word representations in vector space. arXiv preprint arXiv:1301.3781 (2013)

14. Gal, Y., Ghahramani, Z.: A theoretically grounded application of dropout in recurrent neural networks. In: Advances in Neural Information Processing Systems, pp. 1019–1027 (2016)
15. Srivastava, N., Hinton, G.E., Krizhevsky, A., Sutskever, I., Salakhutdinov, R.: Dropout: a simple way to prevent neural networks from overfitting. J. Mach. Learn. Res. **15**(1), 1929–1958 (2014)
16. Aw, A., Mahani, S.A., Lertcheva, N., Kalunsima, S.: TaLAPi - a Thai linguistically annotated corpus for language processing. In: LREC, pp. 125–132 (2014)

Rise of Innovation Applications of Artificial Intelligence

Cardiac Arrhythmia Classification Using Hjorth Descriptors

Thaweesak Yingthawornsuk[✉] and Pawita Temsang

Media Technology Program, King Mongkut's University of Technology Thonburi,
Bangkok, Thailand
thaweesak.yin@kmutt.ac.th, mintchee.mdt@gmail.com

Abstract. The aim of this proposed study is to investigate the discriminant power of Hjorth Descriptor in classification of three categorized groups of subjects' ECG measurement, which are Normal Sinus Rhythm (NSR), Atrial Fibrillation (AF) and Congestive Heart Failure (CHF). This feature has been previously employed to measure the healthiness in persons via their ECG recordings. The algorithm was designed and implemented to extract the Hjorth features and evaluate the performance of classification made on those features by comparing all classifications made among those three databases. Each categorized group included thirty subjects evenly and only three complete QRS complexes of each record in our databases were selected, segmented and extracted for their Hjorth descriptor estimators. In this work three different classifiers were selected, which are Least-Squares (LS), Maximum likelihood (ML) and Support Vector Machine (SVM) for performance evaluation and accuracy comparison. The experimental results from our study showed that the most effective classifier was found to be ML with a mean accuracy of 84.89%, SE of 88.82% and SP of 99.75%, as compared to LS which was found to be the second effective classifier with 88.22% accuracy, and finally SVM with 76.94%. These findings suggested that the promisingly dominant ECG based Hjorth descriptor is capable of class separation among cardiac arrhythmia patient groups.

Keywords: Electrocardiogram · Hjorth descriptor · Classification

1 Introduction

Electrocardiography (ECG) is normally studied by physicists or clinical practitioners for the purpose of heart diagnosis. All electrical activities occur in the heart during measurement will be contained in ECG waveform. How to retrieve the important information from ECG record regarding the person's healthiness level relies highly on an experienced and skillful physicist with a right diagnosing judgement made on ECG monitoring. The time fluctuation and pacing, variation in signal amplitude, frequency response and energy distribution are most common features analyzed from ECG signal and used to determine quantitatively change or significant difference when compared between several different types of ECG signal to normal medical criteria [1].

© Springer Nature Switzerland AG 2019
T. Theeramunkong et al. (Eds.): iSAI-NLP 2017, AISC 807, pp. 95–104, 2019.
https://doi.org/10.1007/978-3-319-94703-7_9

With interestingly attractive and worth to study, ECG signal has been focused and investigated by many research groups attempting to conduct their work on determining the effective way to extract the distinct ECG features which contain the exactly relevant and important information of cardiac condition, and performing the classification for accuracy of class separation when comparative study on multiple classes of ECG signal is acquired [8]. The dynamic time-domain analysis techniques like PCA and AR modeling were successfully applied to ECG signal processing and classification [2, 3]. Principal Component Analysis eliminates the insignificant signal components regarding the number of estimators of Eigen-vector and Eigen-value needed in dimension reduction. Autoregressive modeling provides a way to model signal with model-order coefficients that can represent the time domain characteristic of signal in a form of combination of the ordered model coefficients. The spectral characteristics of ECG signal are another interested topic that has been popularly studied for making an observation on distribution of energy over a specific frequency bandwidth, including frequency response of signal, fundamental frequency and harmonics. The wavelet frequency decomposition was formerly reported as one of the achieving technique of signal processing able to extract information of signal in term of finite energy projected on a continuous family of frequency bands [4].

Recently a group of researchers presented their research work on how to perform the ECG classification based on using a set of variation parameters related to signal amplitude called "Hjorth descriptor". Three categorized groups of ECG signal were classified based on variation of signal amplitude as feature and the result showed the highly successful accuracy of group classification [5, 6]. With less complexity of feature derivation and simplicity of computational implementation, the mentioned feature is re-investigated for its distinguishing characteristic on different groups of ECG signals. The cross-validation scheme was applied for validating the classification performance. The aim of this study is to determine the power of group separation in Hjorth Descriptor features by classifying three different groups of the subjects' ECG waveform. The following sections organized in paper are methodology, experimental results and discussion, and finally conclusion.

2   Methodology

The ECG signals of Atrial Fibrillation (AF) and Congestive Heart Failure (CHF) were collected from MIT-BIH database obtainable online from physionet.org [7]. The ECG measurements of the Normal Sinus Rhythm (NSR) subject group were carried out by using three Bipolar limb leads measurement via BIOPAC MP150 setup on thirty normal female subjects. The ECG signal acquisition was made in all subjects and the same sampling frequency of 1 kHz was used for collecting the signal. Other AF and CHF measurements were obtained at different sampling frequencies of 128 Hz and 250 Hz respectively. In preprocessing state all ECG records in databases were resampled down to 250 Hz so that the frequency response of all types of ECG signals can be compared each other in the same frequency range of 0–125 Hz. Each group of ECG signals evenly consists of 30 records and only three complete QRS complexes were taken from individual ECG records for a window length of 2–3 s for analysis and feature extraction in further states. Figure 1 shows a whole procedure of Hjorth extraction from the imported ECG signal file by following all calculations on Eqs. 1 and 2, and parameter descriptions in the following details.

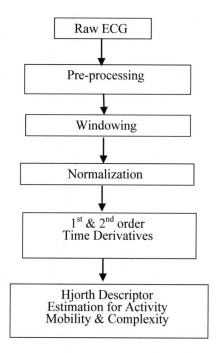

Fig. 1. Procedure of Hjorth Extraction on the raw ECG signal

The NSR is normal condition of Electrocardiography which is composed of a P wave, a QRS complex, and a T wave (see Fig. 2). The AF is an irregular Electrocardiography with no P wave (see Fig. 3) because of numerous small depolarization waves spread in all direction through the atria. These waves are weak and many of them are of opposite polarity. Therefore, at given time they are completely neutralized each other. The CHF is a chronic progressive condition that affects the pumping power of heart muscles. This refers to the state in which fluid builds up around the heart and causes it to pump inefficiently.

Before feature extraction was conducted, we first made observation on how spectra of ECG signals respond in frequency domain by estimating Power Spectral Density (PSD) of all three typical ECG signals in database. The significant difference in frequency bandwidth was clearly notified between NSR and CHF signals by looking at the most concentrated spectral power located on a specific frequency range. As one can see in Figs. 2 and 4, the CHF spectrum had most power distributed over a very narrow frequency range of 0–20 Hz while the power of NSR distributed over a wider frequency bandwidth.

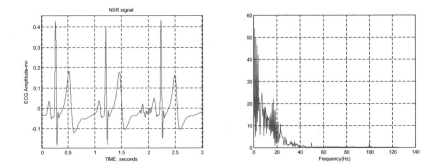

Fig. 2. ECG signal of NSR subject and its PSD

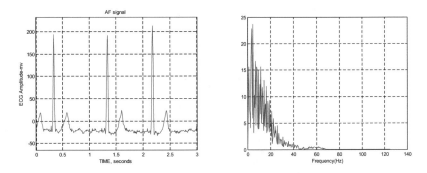

Fig. 3. AF ECG signal and its PSD

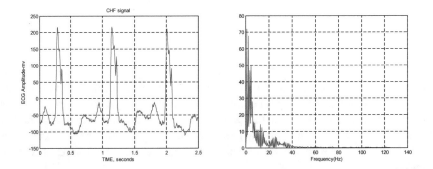

Fig. 4. CHF ECG signal and its PSD

Hjorth descriptor is defined as a set of parameters that can represent the measure of ventricle repolarization on ECG waveform. Hjorth descriptor consists of Activity, Mobility, and Complexity. The following steps show how to estimate Hjorth by assuming that $x(n)$ is a ECG signal with $n = 0, 1, 2, ..., N-1$. And $x(n)'$ is defined as the 1st-order Time derivative derived from the following sequence signal,

$$X(n)' = x(n) - x(n-1), \quad n = 1, 2 ..., N \tag{1}$$

And $x(n)''$ is defined as 2nd-order time derivative,

$$x(n)'' = x(n)' - x(n-1)' \tag{2}$$

The activity parameter can be derived from σ_x, standard deviation of $x(n)$ and furthermore σ_x' and σ_x'' are standard deviations of $x(n)'$ and $x(n)''$, respectively.

Activity is defined as deviation quadratic parameter which is a variance of signal, σ_x^2. Mobility is defined as a ratio of standard deviation, $\dfrac{\sigma_{x'}}{\sigma_x}$. Complexity is defined as a ratio of Mobility parameters, $\dfrac{\sigma_{x''}/\sigma_x}{\sigma_x/\sigma_x}$.

To determine the performance of extracted Hjorth features in classifying among typical ECG groups, several classifiers were chosen for comparison, which consists of Maximum likelihood (ML), Support Vector Machines (SVM) and Least Square (LS). The Cross Validation (CV) technique was applied to each classifier to perform 100 folds and to obtain a mean value of correct classification scores from all folds. In this study, all classifications were performed in pairwise against each other between two ECG classes. In this way, the sensitivity (S.E.) and specificity (S.P.) can be calculated directly from true positive and true negative which are class-1 data identified by classifier correctly as data of class 1 and class 2 data identified correctly as class 2 data. The data prior to classifications were separated into two portions served as inputs to two states of classification, one for training a classifier and second for testing that same classifier for final performance evaluation and the 100-fold CV technique was included in these states as well.

3 Experimental Results and Discussion

The activity, mobility and complexity extracted from all typical ECG signals showed the different tendency along three ECG classes. All parameters were presented in their mean values depicted in Figs. 5, 6 and 7. The activity of CHF shows as the highest quantity with respect to those of AF and NSR. In Fig. 5 mobility and complexity values seems to be very small compared to activity. Inversely shown in Fig. 6, the mobility of NSR is the highest value and very different from those of CHF and AF respectively. There is no different mobility value between CHF and AF. The tendency of quantity change as similar to a case of activity can be found in complexity with the lowest complexity value in NSR case (see Fig. 7). As seen in comparisons the variations of Hjorth descriptor can be used to measure the quantitative difference among classes.

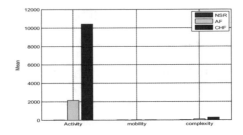

Fig. 5. Quantitative comparison among three Hjorth parameters extracted from all three ECG types of NSR, AF and CHF

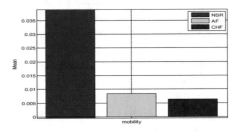

Fig. 6. Mean values of Mobility feature among three different ECG types

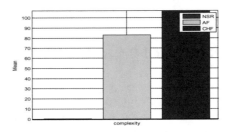

Fig. 7. Mean values of Complexity feature among three different ECG types

All results obtained from classification were combined in Table 1. The percentages of correct classification, SE and SP can reach a 100 in NSR-AF pairwise for both classification using ML and SVM. The worst case scenario also occurred in classifying NSR and AF with ML and CHF against other two with SVM. Table 1 shows summarized results of applying PCA feature dimension reduction prior to performing all classifications.

Table 1. Summarized mean values of SE, SP and Accuracy from pairwise classifications with PCA preprocessing

Pairwise		SE	SP	%ACC
LS	NSR-AF	100	99.95	97.25
	NSR-CHF	99.97	99.64	85.46
	AF-CHF	99.90	99.18	71.76
ML	NSR-AF	100	100	100
	NSR-CHF	N/A	100	50
	AF-CHF	99.83	99.59	77.88
SVM	NSR-AF	100	100.	100
	NSR-CHF	N/A	100	50.21
	AF-CHF	N/A	100	49.88

Figure 8 presents the comparative plots of all accuracies from Table 1. Surprisingly, classifying NSR and AF with all LS, ML and SVM is highly effective and the LS classifier provides high accuracies for all pairwise cases, including the highest SE and SP values of 100%. These findings can be concluded that all three parameters of Hjorth descriptor are capable of separating different classes of ECG signal. Further study will be more carried out on modification of signal preprocessing regarding window length of signal analysis and filtering. Furthermore, the derivation of higher order of Hjorth descriptor should be also taken in account to gain more improving power of class discrimination.

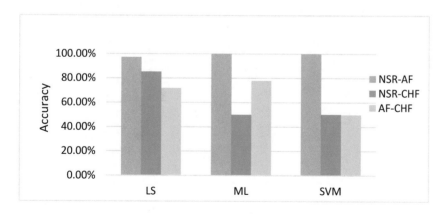

Fig. 8. Comparison of mean values of classification accuracy with PCA

Another development of GUI was also implemented on the benefit of clinical level use. However, not all matlab scripts that we developed and used to achieve the aim of this proposed work now can be completely executed in GUI. Carrying on the development of GUI would be included as one of our goals and expansion of this research study in the future direction.

The screen captures of GUI presented in Figs. 9, 10, 11, 12 and 13 are the displays of the plots of: Raw ECG signal after loading ECG files by clicking on a designed drop bar on the top of window popup, segmenting signal by pressing "Plot Segmentation" button, estimating PSD of imported ECG signal by clicking "Plot Spectrum" button, and calculating Activity, Mobility and Complexity values by pressing "Hjorth Descriptor" button. Finally, two-dimension scatter plot of all ECG classes by selecting pairwise features in a drop bar and then clicking a "Plot" button. The distribution of feature data can be explored and observed visually first to gain more information on judgement that would be helpful before doing classification and further steps.

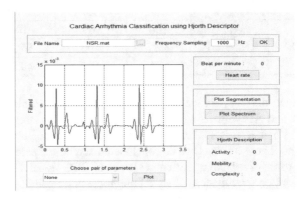

Fig. 9. GUI development for friendly use in ECG signal segmentation

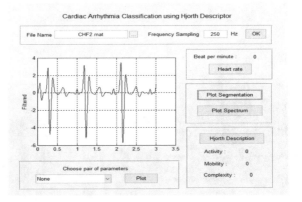

Fig. 10. Signal segmentation made on CHF waveform

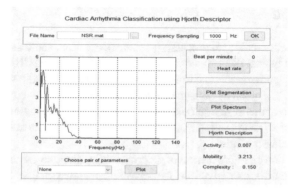

Fig. 11. Estimated PSD and Hjorth feature values from NSR signal

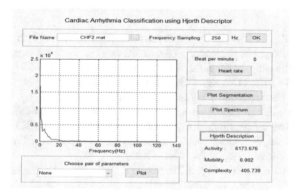

Fig. 12. Estimated PSD and Hjorth feature values from CHF signal

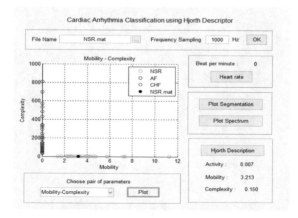

Fig. 13. Scatter plot of Mobility vs Complexity estimated from three typical ECG signal classes with an additive NSR imported, extracted, and identified as a dark dot

4 Conclusion

This work concludes the signal processing, feature extraction, and feature classification of three different typical ECG signals acquired from NSR, AF and CHF subjects. Based on results presented here, the maximum accuracy can reach a 100% with LS, ML and SVM classifiers. The main advantage of Hjorth descriptor is simplicity of feature derivation and less computational implementation but the output features are oppositely very effective in terms of representing as powerful class discriminators and achieving high performance of classification.

Further direction on improvement of preprocessing techniques, signal segmentation, noise filtering would be carried out and the derivation of higher order of Hjorth descriptor should be also taken in account of more investigation to achieve practical use.

References

1. Brosche, T.A.M.: The EKG Handbook. Jones & Bartlett Publisher (2010)
2. Bollmann, A., Roig, M., Castells, F., Laguna, P., Leif, S.: Principal component analysis in ECG signal processing. EURASIP J. Adv. Signal Process. **2007**, 074580 (2007)
3. Xiao, Q.U., Wei, C.: ECG signal classification based on BPNN. In: 2011 International Conference on Electric Information and Control Engineering (2011)
4. Daamouche, A., Hamami, L., Alajlan, N., Melgani, F.: A wavelet optimization approach for ECG signal classification. Biomed. Signal Process. Control **7**(4), 342–349 (2012)
5. Hadiyoso, S., Rizal, A.: ECG Signal Classification using Higher-Order Complexity of Hjorth Descriptor. American Scientific Publishers (2015)
6. Rizal, A., Hadiyoso, S.: ECG Signal Classification Using Hjorth Descriptor
7. Physionet.org, "ECG Database". http://physionet.org/physiobank/database/#ecg
8. Tompkins, W.J.: Electrocardiography. In: Tompkins, W.J. (ed.) Biomedical Digital Signal Processing. Prentice Hall, New Jersey, pp. 24–54 (2000)
9. Guyton, C: Textbook of Medical Physiology, 8th edn. Harcourt College Pub., October 1990

CCTV Face Detection Criminals and Tracking System Using Data Analysis Algorithm

Patiyuth Pramkeaw[1(✉)], Pearlrada Ngamrungsiri[1],
and Mahasak Ketcham[2]

[1] Department of Media Technology,
King Mongkut's University of Technology Thonburi, Bangkok, Thailand
patiyuth.pra@kmutt.ac.th, kaimookmpn@gmail.com
[2] Faculty of Information Technology,
King Mongkut's University of Technology North Bangkok, Bangkok, Thailand
mahasak.k@it.kmutnb.ac.th

Abstract. The research aimed to the study the development of CCTV face detection criminals and tracking system using data analysis algorithm. The proposed algorithm reduce the time spent searching for criminals and finding suspicious persons or criminals in society and precision with technology applied. It can recognize many faces. This research utilized the CCTV images to be analyzed by face detection technique. By sending a notification via text message and email. Results from Single Face Detection and group face detection were obtained as comparison. The accuracy of program was 91%. The system can recognize many faces and can be used to secure the place.

Keywords: CCTV · Face detection · Image processing

1 Introduction

Crime can destroy the life or property of the public and making people to be fear. Crime was committed by individuals or groups of persons known as criminals, which society nowadays tend to be increased by the advancement of technology and the modernization of communication, especially in public places where people are busy with themselves such as the bombing incident at Thao Maha Phrom Shrine Ratchaprasong Intersection or bombing in the economic district and several provinces of southern have so many innocent people injured and got killed. Some cases, the criminals were shill not arrested and prosecuted. This unsecure situation affected to the daily life of the people, tourism industry in Thailand and impacted the overall of country economy.

At present Image Processing Technology, is the process of managing and analyzing information of images by computer evaluating for example image conversion, image quality improvement, face detection and CCTV technology, are continuously improving and widely used [1].

As a result of the increasing crimes and fugitives this can bring to program development by analyzing of CCTV images with use of face detection technique. This will help police officers and investigators finding the offenders and bring them to justice faster and more efficient.

© Springer Nature Switzerland AG 2019
T. Theeramunkong et al. (Eds.): iSAI-NLP 2017, AISC 807, pp. 105–110, 2019.
https://doi.org/10.1007/978-3-319-94703-7_10

2 Related Background

2.1 Face Detection

Face Detection is the process of finding person face from pictures or videos, then processing face by using the algorithm is to detect the face. For example of the algorithm used to detect the face is the Viola-Jones object detection framework was presented by Marks and Jones in 2015 [2, 3]. Three steps of Viola-Jones' algorithm are as followings:

2.1.1. Integral Image by using Haar-like model to detect the face of Viola-Jones, consisting of two rectangular areas are shading area and not shading area. Finding Haar-like model determined by the difference in light intensity between shading and not shading. When the results are compared with the threshold value and the polarity used to determine whether the input image is a face image or not. The results will be taken into consideration in the next steps.

2.1.2. AdaBoost is the process for finding features which are similar and different to the image, the process is as follows: Define weight for Haar-like feature, use Haarlike feature to check on image and separate the desired image and undesired.

2.1.3. The process of combining the clusters is a technique used to increase the detection efficiency for accuracy and decreasing computation time.

2.2 Face Recognition

Face Recognition is the process of detecting facial image from the face detection process to compare with the facial database. An example of facial recognition algorithms, Eigen face [4, 5], is the extraction features of the face by using reduction of information dimension. The procedure is divided into 5 steps: input and average image, find image difference, calculate covariance matrix, calculate eigenvector and weight vector [6].

3 Literature Review

"The Application of Scale Invariant Feature Transform fused with Shape Model in The Human Face Recognition", which is a research work aims to develop a procedure of human face recognition using Eigen face. In addition, this research will bring the image from the learning group to create the Eigen face by applying the principal component analysis theory of straight face images, then calculating the weight to represent or model of learning group [7].

"A Face Detection Algorithm Based on Adaboost and New Haar-Like Feature", this research aims to develop a user control for facial recognition and for program developers to implement the program's security enhancement and face detection by Haar-like features. Facial recognition and PCA features are main operation of this control consisting of two stages, learning and implementation [8].

"Towards Robust Face Recognition for Intelligent-CCTV based Surveillance using One Gallery Image", present the investigation of the human face and motion via CCTV following the movement of people inside the building by using face detection technology. There are three parts used in the system: Human detection, Face detection and Face recognition [9].

4 Development Methods and Techniques Used

4.1 System Operation

See Fig. 1.

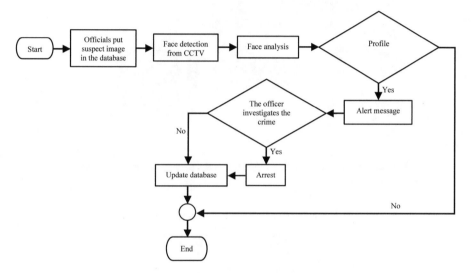

Fig. 1. System workflow

4.2 Arrangement of the Suspect S' Face Database

The officer must add criminals face database, which must be straight faces and they can be gray or color images into database to compare with the faces detected by the CCTV cameras.

4.3 Face Detection on CCTV

The image from the video camera is analyzed for the position of the face, then calculating each section of the face by using an algorithm Haar-Like to find the face, eyes, ears, nose and mouth, which can be used to identify face.

4.4 Face Analysis

From the face detection phase, indicates what part of the image in the CCTV is the facial image. Analysis with face recognition using the Eigen face algorithm to compare faces detected from the CCTV to the criminal face in the database.

4.5   Notification of Matched Image

When a face image from CCTV was found to match with criminal face in database, alerting message or email will be sent to relevant person. In order for the staff or related persons in the vicinity to check in after the inspection has been completed, the staff should update the database to the present (Figs. 2, 3 and 4).

Fig. 2. The face recognition algorithm can analyze or identify individual

Table 1. Accuracy of single face detection

NO	Detection results	% Accuracy
1	Detect success	95%
2	Detect success	94%
3	Detect success	94%
4	Detect success	95%
5	Detect success	95%

Fig. 3. Experiment with groups

Table 2. Accuracy of multi face detection

NO	Detection results	% Accuracy
1	Detect success	91%
2	Detect success	91%
3	Detect success	93%
4	Detect success	94%
5	Detect success	92%

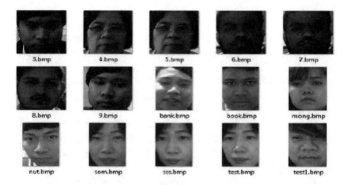

Fig. 4. The image result from the experimental database

4.6   Experiment of Face Detection and Face Analysis

In this section, the researchers have studied the program to be applied for face detection and facial analysis. The experiment will use CCTV input image of criminals face through the process of face detection process. Then facial analysis has used with facial recognition process to compare with stored faces in database (Tables 1 and 2).

5   Conclusion

Experimental face analysis is able to identify individuals' for successfully. There are 10 images in the database used in this study. In five times tests, the results showed that all five trials both individual face and multiple face detection are more accurate than 90%. Error in detection was found in this study because of tilling position of face in image.

Acknowledgments. This research has been financially granted by the National Research Council of Thailand and Department of Media Technology at King Mongkut's University of Technology Thonburi. This paper presented the result of research study corresponding to the research project id number: 347920 approved by National Research Council of Thailand.

References

1. Manlises, O., Jesus, M., Jackson, L., Czarleine, K.: Real-time integrated CCTV using face and pedestrian detection image processing algorithm for automatic traffic light transitions. In: IEEE, pp. 44–77 (2015)
2. Ahmed, E., Jones, M., Marks, T.K.: An improved deep learning architecture for person re-identification. In: IEEE, pp. 3908–3916 (2015)
3. Akshay, A., Marks, T.K., Jones, M.J., Tieu, K.H., Rohith, M.V.: Fully automatic pose-invariant face recognition via 3D pose normalization. In: IEEE International Conference on Computer Vision, pp. 937–944 (2011)
4. Kshirsagar, V.P., Baviskar, M.R., Gaikwad, M.E.: Face recognition using eigenfaces. In: IEEE (2011)
5. Xu, Y., Li, Z., Zhang, D.: A survey of dictionary learning algorithms for face recognition. In: IEEE Translation and Content Mining are Permitted for Academic Research only, pp. 1–12 (2016)
6. Patiyuth, P.: The study analysis knee angle of color set detection using image processing technique. In: International Conference on Signal-Image Technology & Internet Based System, pp. 657–660. IEEE (2016)
7. Jing, K., Wu, F., Zhu, X., Dong, X., Ma, F., Li, Z.: Multi-spectral low-rank structured dictionary learning for face recognition. Pattern Recognit. **59**, 14–25 (2016)
8. Songyan, M., Bai, L.: A face detection algorithm based on Adaboost and new HaarLike feature. In: IEEE, pp. 651–654 (2016)
9. Ting, S., Shaokang, C., Conrad, S., Brian, C.: Towards robust face recognition for intelligent-CCTV based surveillance using one gallery image. In: IEEE, pp. 470–475 (2007)

Isan Dhamma Characters Segmentation and Reading in Thai

Siriya Phattarachairawee[✉], Montean Rattanasiriwongwut,
and Mahasak Ketcham

Faculty of Information Technology, King Mongkut's University of Technology
North Bangkok, Bangkok, Thailand
william.siriya@gmail.com, {montean,mahasak.k}@it.
kmutnb.ac.th

Abstract. This paper presents Isan Dhamma Characters Segmentation and Reading in Thai, Palm leaf manuscript is considered as a kind of cultural heritage and the record of local wisdom of ancestors that should be transformed into digital format for educational and research benefits of the next generation. This research presents palm leaf manuscript's Isan Dhamma characters segmentation and reading conducted by using image processing. The objective of this research is to utilize the obtained data in sentence recognition process further. The input was digital photos of a palm leaf manuscript written with Isan Dhamma characters that was proposed to be adjusted on its quality by adjusting light intensity through histogram. Subsequently, its quality was improved by using median filter in order to screen data on enhancement or attenuation of some picture's properties in order to gain quality as demanded. Subsequently, characters were segmented from background (segmentation) by using Global Thresholding. In the last process, each character was recognized by using the principles of neural network and compared with support vector machine. After conducting the experiment with 10 images of palm leaf manuscript, it was found that neural network gives better effects than support vector machine by 98%.

Keywords: K-NN · Threshold · Morphology · CM calculation

1 Introduction

Digital Age is originated from technological advancement including computer technology, communication technology, and information technology. Consequently, the growth rate of information has been changed rapidly. The format of information would be in message, image, and sound formats and such information has been collected in digital formats. Consequently, it is convenient for retrieval and application as demanded. However, there are some types of information that are collected in document format without transforming to digital format therefore it is difficult for retrieval and practical utilization. Consequently, knowledge on image processing is adjusted in order to adjust document to be digital image and operated with Optical Character Recognition (OCR) for transforming digital image to be digital document. To transform

© Springer Nature Switzerland AG 2019
T. Theeramunkong et al. (Eds.): iSAI-NLP 2017, AISC 807, pp. 111–118, 2019.
https://doi.org/10.1007/978-3-319-94703-7_11

digital images to be in electronic document format, it is necessary to utilize knowledge on image processing and Optical Character Recognition (OCR).

Thai character recognition can be applied to several businesses, for example, news reporters are able to record all interviews on paper or students are able to summarize their study on notebooks. Subsequently, they input such paper into OPCR system and the system will transform characters to be in the form of character process program. Consequently, it is not necessary for news reporters and students to type all characters repetitively. This helps to reduce and accelerate work process. Thai handwriting character recognition program can be applied to many devices, for example, electronic whiteboard or PDA for online operation, for example, when writing on electronic whiteboard, program will be able to be transformed in the form of character processing program leading to higher operational efficiency.

Generally, [18, 19] the principles of Optical Character Recognition are different upon the target languages because each language has distinctive format. For recognition technology, Dhamma characters have been developed to support recognition of various sizes and fonts of characters in order to improve accuracy on recognition from Thai language directly, especially in the form of similar Dhamma characters. Researches related to Dhamma character recognition proposed the learning methods based on various characters, for example, Back-Propogation Neuron Network (BPN), Time-Delay Neuron Network (TDNN), or FuzzyLogic, etc.

AbdelRaouf et al. [1] presented an extensive study and analysis on several verbs of Corpus Arabic language that was suitable for using in ORC developed by artificial neural networks used in segmenting among different Arabic characters automatically. Oujaoura et al. [2] developed OCR system of Arabic segmentation and typed Zernike of the period spent in the period with no transformation of Walz and extraction process as well as artificial neural networks. Abulnaja and Batawi [3] used N-Program Writing Techniques and fault techniques to enhance correctness of Arabic language. Moussaet al [4] used this program to analyze surface in order to reduce difficulties in processing of recognition system and to obtain successful Arabic character printing recognition. Patel et al. [5] identified the problems of handwriting character recognition enhanced by Multi-resolution using DWT and Euclidean distance. Patel et al. [6] conducted this research to recognize characters in scanned and defined documents as well as to study on effects of changing of artificial neural networks. Abed et al. [7] proposed how to reduce complexity of handwriting characters based on recognition on behavioral imitation of birds that was called Optimization (PSOA). Sahu et al. [8] explained classification based on learning from extensively accepted samples for recognizing characters from 1990.

The research is divided into 5 parts and the second part is explanation on standards and procedures/methods used on paper for enhancing efficiency and segmenting characters while the third part is explanation on preparation of data presentation. The forth part is explanation on recognition methods while the fifth part is explanation on results and conclusion.

2 The Proposed Method

In this paper, we proposed the methods for isan dhamma characters segmentation and reading in thai, including morphology, preparation of image data and recognition.

2.1 Median Filter [9–11]

Firstly, images are enhanced by using mean filtering through spatial convolution with the weight of 1 thoroughly. After conducting spatial convolution, the outcome is averaged by selecting mean as shown in the equation below:

$$g(i,j) = med\left[x(i+r, j+s) \in A(j,j) \in za^2\right] \tag{1}$$

Whereas, mean filtering is a kind of non-linear filtering and the advantage of image enhancement through mean filtering is image blurring giving the same tone of pixel's intensity.

2.2 Morphology [13, 14]

In the following procedure, noises were removed by using image corrosion technique in order to decrease the area of white color or object in the image. The processing procedures were the same as those of image expansion but there was difference based on the principles stated that all positions with factors moving would be compared with images.

$$A \ominus B = \left\{z|(B)_2 \subseteq A\right\} \tag{2}$$

where A the image requiring corrosion of image border;
   B structural elements;
   Z pixel field (Fig. 1).

Fig. 1. Image after enhancing with morphology

3   The Proposed Method

See Fig. 2.

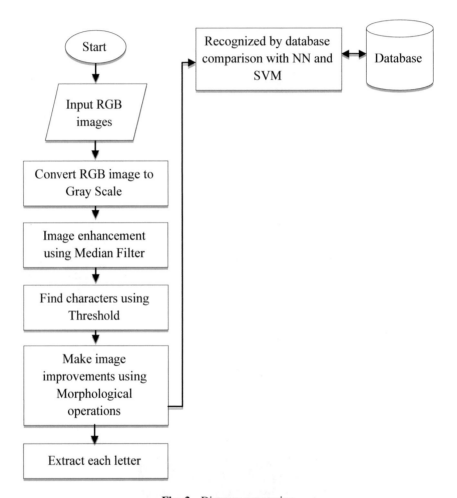

Fig. 2. Diagram processing

3.1   Preparation of Image Data

This research presented the method for collecting data obtained from stored documents in the form of image file of 50 images of the characters used as the sentences and its file name extension was Joint Photographic Experts Group (JPEG) with the sizes ranged from 320 × 240 to 2048 × 1536 pixels with white characters and black background (Figs. 3 and 4).

3.2   Recognition

Artificial Neural Networks [16, 17]. Learning of Perceptron network is in the manner of layered structure and learning data will be input in the first layer that is Input Layer. After calculating the first layer, it will be sent to the Hidden Layer whereas each unit of this layer will accept data from all units in the former layers for calculating and sending to the next layer. When data is sent to the last layer or Output Layer, the input will be given by the system. This kind of data transmission is considered as feed forward. Subsequently, the output will be checked by the system how it is incorrect based on the target.

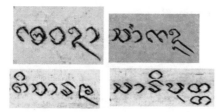

Fig. 3. Sample images used in testing

พยัญชนะวรรค ก

อักษรไทย	ก	ข	ค	ฆ	ง
อักษรธรรม	ꪶ	ꪎ	ꪀ	ꪑ	ꪊ

พยัญชนะวรรค จ

อักษรไทย	จ	ฉ	ช,ฌ	ญ	ฌ
อักษรธรรม	ꪒ	ꪓ	ꪔ	ꪕ	ꪖ

พยัญชนะวรรค ฏ

อักษรไทย	ฏ	ฐ	ฑ	ฒ	ณ
อักษรธรรม	ꪗ	ꪘ	ꪙ	ꪚ	ꪛ

พยัญชนะวรรค ต

อักษรไทย	ต	ถ	ท	ธ	น	บ
อักษรธรรม	ꪚ	ꪜ	ꪝ	ꪞ	ꪟ	ꪠ

พยัญชนะวรรค ป

อักษรไทย	ป,บ	ผ	ฝ	พ	ฟ	ภ	ม
อักษรธรรม	ꪡ	ꪢ	ꪣ	ꪤ	ꪥ	ꪦ	ꪧ

พยัญชนะเศษวรรค

อักษรไทย	ย	ร	ล	ว	ศ,ษ,ส
อักษรธรรม	ꪨ	ꪩ	ꪪ	ꪫ	ꪬ

อักษรไทย	ห	ฬ	อ	ฮ
อักษรธรรม	ꪭ	ꪮ	ꪯ	ꪰ

Fig. 4. Full character that was written in full form upon normal format with 37 characters

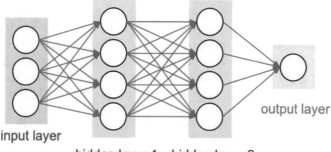

Fig. 5. Layered artificial neural networks used in backpropagation instruction

In this research, we utilized instructional algorithm for using with artificial neural networks based on Backpropagation technique. This technique is the use of layered structure of Supervised Learning with demanded target and adaptive networks were used for adjusting the weight properly as shown in Fig. 5.

It is operated by multiplying data with the weight of each leg. All results of data from all legs will be combined and compared with the target value to have the most similar value. The important thing is to perceiving the weight that is uncertain but it can be adjusted by programing computer with Levenberg-Marquardt Algorithm Learning that is classified as the methodology used in teaching the fastest artificial neural networks. However, there is some complex calculation requiring more space of computer's memory unit. The operational process of Levenberg-Marquardt Algorithm can be explained as follows:

Calculate (W) based on Eq. 3:

$$(W) = \sum_{p=l}^{p} \left(d_{lp} - O_{lp} \right) 2 \tag{3}$$

Calculate Jacobian Matrix based on Eq. 4.

$$J = \begin{bmatrix} \frac{\partial e_{11}}{\partial w_1} & \frac{\partial e_{11}}{\partial w_2} & \Lambda & \frac{\partial e_{11}}{\partial w_N} \\ \frac{\partial e_{21}}{\partial w_1} & \frac{\partial e_{21}}{\partial w_2} & \Lambda & \frac{\partial e_{21}}{\partial w_N} \\ M & & \Lambda & M \\ \frac{\partial e_{K1}}{\partial w_1} & \frac{\partial e_{K1}}{\partial w_2} & \Lambda & \frac{\partial e_{K1}}{\partial w_N} \\ M & & \Lambda & M \\ \frac{\partial e_{1P}}{\partial w_1} & \frac{\partial e_{1P}}{\partial w_2} & \Lambda & \frac{\partial e_{1P}}{\partial w_N} \\ M & & \Lambda & M \\ \frac{\partial e_{KP}}{\partial w_1} & \frac{\partial e_{KP}}{\partial w_2} & \Lambda & \frac{\partial e_{KP}}{\partial w_N} \end{bmatrix} \tag{4}$$

Calculate weighting +1 by solving Eq. 5.

$$W_{t+1} = W_t - \frac{1}{\mu_t} \left[I - \frac{\hat{J}_t^T \hat{J}_t}{\mu_t + J_t^T \hat{J}_t} \right] \left(J_t^T \hat{E}_t \right) \tag{5}$$

Calculate new F(W) for comparing with exiting value. If new F(W) is less than exiting value, it must be divided with v and back to the operation at 1. If the value of new F(W) is not reduced from existing value, it must be multiplied with v and back to the operation at 3. At the start, μvalue is often a small value, for example, μ = 0.01 and v is the value applied to the value of μ during each operation whereas v > 0, for example, v = 10 by using input layer as 1, Hidden layer as 3, and output layer as 1.

4 Experimental Results

This research developed the program by using Matlab and 50 sample images of palm leaf manuscript's Isan Dhamma characters consisted of 550 characters and 120 digits therefore the total characters were 670. Table 1 represents the result on accuracy compared between SVM method and artificial neural networks (Figs. 6, 7 and 8).

Table 1. The result on accuracy compared between SVM method and artificial neural networks

Method	Accuracy rate
Support vector machine	96%
Neural network	98%

Fig. 6. Image of character segmentation

Fig. 7. Demonstration of recognition with correct spelling

Fig. 8. Demonstration of recognition with incorrect spelling

5 Conclusion

The proposed guidelines on palm leaf manuscript's Isan Dhamma characters segmentation and reading was started from image enhancement by using histogram and median filter. Subsequently, the obtained images were threshed by segmenting images in order to enable each image segment to use suitable threshold. However, there were some side-effects increasing noises therefore it was necessary to choose proper block size and different block sizes varied upon images. Finally, recognition must be operated by using the principles of neural network for comparing with support vector machine.

References

1. AbdelRaouf, A., Higgins, C.A., Pridmore, T., Khalil, M.: Building a multi-modal Arabic corpus (MMAC). Int. J. Doc. Anal. Recognit. (IJDAR) **13**(4), 285–302 (2010)
2. Oujaoura, M., El Ayachi, R., Fakir, M., Bouikhalene, B., Minaoui, B.: Zernike moments and neural networks for recognition of isolated Arabic characters. Int. J. Comput. Eng. Sci. (IJCES) **2**(3), 17–25 (2012)
3. Abulnaja, O., Batawi, Y.: Improving Arabic optical character recognition accuracy using N-version programming technique. Can. J. Image Process. Comput. Vis. **3**, 44–46 (2012)
4. Moussa, S.B., Zahour, A., Benabdelhafid, A., Alimi, A.M.: New features using fractal multi-dimensions for generalized Arabic font recognition. Pattern Recognit. Lett. **31**(5), 361–371 (2010)
5. Patel, C.I., Patel, R., Patel, P.: Handwritten character recognition using neural network. Int. J. Sci. Eng. Res. **2**(5), 1–6 (2011)
6. Patel, D.K., Som, T., Yadav, S.K., Singh, M.K.: Handwritten character recognition using multiresolution technique and euclidean distance metric (2012)
7. Sahu, V.L., Kubde, B.: Offline handwritten character recognition techniques using neural network: A review. Int. J. Sci. Res. (IJSR) **2**(1), 87–94 (2013)
8. Abed, M.A., Alasadi, H.A.A.: Simplifying handwritten characters recognition using a particle swarm optimization approach. Eur. Acad. Res. **1** (2013)
9. Mohammed, J.R.: An improved median filter based on efficient noise detection for high quality image restoration. In: Second Asia International Conference on Modeling and Simulation, AICMS 2008, pp. 327–331. IEEE Press (2007)
10. Pahsa, A.: Morphological image processing with fuzzy logic. J. Aeronaut. Space Technol. **2**(3), 27–34 (2006)
11. Wayne, L.W.C.: Mathematical morphology and its applications on image segmentation. Department of Computer Science and Information Engineering, National Taiwan University (2000)
12. Poggio, T., Girosi, F.: Networks for approximation and learning. Proc. IEEE **78**(9), 1481–1497 (1990)
13. Chamchong, R., Fung, C.C.: Text line extraction using adaptive partial projection for palm leaf manuscripts from Thailand. In: 2012 International Conference on Frontiers in Handwriting Recognition (ICFHR), pp. 588–593 (2012)
14. Khakham, P., Chumuang, N., Ketcham, M.: Isan Dhamma handwritten characters recognition system by using functional trees classifier. In: 11th International Conference on Signal-Image Technology and Internet Based Systems (SITIS), pp. 606–612 (2015)
15. Chumuang, N., Khakham, P., Ketcham, M.: The intelligence algorithm for character recognition on palm leaf manuscript. Far East J. Math. Sci. **98**(3), 333 (2015)

Ontology-Assisted Structural Design Flaw Detection of Object-Oriented Software

Sakorn Mekruksavanich[✉]

Department of Computer Engineering, School of Information and
Communication Technology, University of Phayao, Phayao, Thailand
sakorn.me@up.ac.th

Abstract. A design flaw is indicative of a potential shortcoming in the
construction of a software system and can result in a reduction in the quality of
the software. This study offers a means of detecting design flaws through the use
of ontology-assisted flaw description and declarative meta programming.
Ontology flaw structures are used to describe the flaw domains, while the use of
a declarative-based method allows the design flaws which arise within an object-
oriented system to be altered into these structures, thereby permitting their
detection at the metalevel when declarative meta programming is employed.
This research uses the method described in order to detect a number of design
flaws which are already well-documented. The findings demonstrate that the
method is successful in detecting those flaws, and that structural design flaw
detection is particularly effective.

Keywords: Design flaws · Detection · Object-oriented design
Ontology

1 Introduction

One of the latest frameworks which can be effectively adapted to develop information
systems is the object-oriented paradigm. This offers a much more suitable approach
than the traditional procedural code [2]. Its advantages lie in the fact that it gives the
developer a set of tools by which complexity can be managed. These include data
abstraction, modularity, encapsulation, polymorphism, and inheritance. The typical
objectives of software development, such as ease of maintenance and re-use can
therefore be readily supported by object-oriented programming [3], but the challenge in
its implementation is that developers require considerable experience and expertise to
realize the benefits of this object-oriented paradigm. When inexperienced programmers
are involved, their errors can lead to lower quality software, even though it may appear
functionally adequate. Design flaws are a common problem in such situations where
software is of lower quality.

Design flaws are program properties that show the potential erroneous design of a
software system and are introduced in the early stages of software development or
during evolution. These flaws may result in low reuse, high complexity, and low
maintainability in developed systems. In the recent literature, design flaws are referred
to as Bad Smells [3] and AntiPatterns [1]. Many authors normally denote design flaws

T. Theeramunkong et al. (Eds.): iSAI-NLP 2017, AISC 807, pp. 119–128, 2019.
https://doi.org/10.1007/978-3-319-94703-7_12

by using metaphors. They propose many approaches on how to recognize and correct such erroneous software flaws. Accordingly, it is necessary for the modern development of software to incorporate features that can be used to identify and remove flaws in a design. The stages in the software development life cycle, in which the identification and elimination of the design flaws are generally practiced, include the development, testing, and maintenance stages. One of the more popular techniques that is used to process these flaws is called refactoring. During the process, the internal design of the system is altered, while the observable behavior is maintained. The refactoring technique is considered an important activity, especially in agile methodologies.

The identification of design flaws in the early age comes via manual detection methods such as the code inspection technique [11]. This technique involves carefully checking source code, examining the design and documentation of software, and checking for potential problems based on past experience. However, this approach is time-consuming, non-repeatable, and non-scalable, and these are considerable disadvantages [6]. To avoid the limitations of those issues, heuristic metrics are proposed to identify design flaws in software systems. The detection of software metrics uses a predefined set of thresholds to interpret the detection results. Although effective in flaw detection, metric-based flaw identification depends inevitably on proper metrics and thresholds which are used to detect the flaws. The question of how to obtain the optimal relationship between software metric properties and design flaws has been the focus of much research.

The goal of this paper is to propose a new ontology-assisted approach for design flaw detection. In this study, the model of ontology semantic-based meaning for software design flaws and their detection is presented. Two groups of processes of flaw detection can be employed, namely the Representation and Detection phases. The first process performs the explicit description of flaw concepts and the proper rule selection for such flaw detection. The second process performs flaw predictions and shows the results of the design flaw detection. Two open source software types are used for evaluating the performance of the proposed approach. The experimental results show that the proposed scheme can establish suitable design flaw detection. Moreover, it is also simple and saves time because its implementation does not require expert knowledge as would be the case with the traditional metric-based detection approach. The proposed methodology can provide software developers and project managers with dependable indicators of design flaws in software systems.

This paper is organized as follows. Section 2 describes some knowledge involved with the research. The proposed methodology is discussed elaborately in Sect. 3. Section 4 presents the experimental studies including experimental results. Conclusions and future works are finally described in Sect. 5.

2 Theoretical Background

2.1 Design Flaws

Design flaws in the context of object-oriented programming are first introduced by Websters book [10]; the book contributes to conceptual, political, coding, and quality-assurance problems. Riel [8] defined some groups of heuristics characterizing good object-oriented programming to manually assess system quality for improving its design and implementation. Fowler [3] proposed 22 low-level code smells. These smells are in design problems in source code and they should be removed by refactoring. These smells are described in an informal way and need interpretation with a manual detection process. The classification of design flaws is proposed in many works [4]. These taxonomies are introduced to provide a better understanding of the smells and to recognize the relationships between smells. By these classifications, researchers can improve the precision of design flaw detection. Brown et al. [1] define 40 antipatterns which are described in depth in terms of code smells-like descriptions. They are the basis of all approaches to specify and to detect code smells semi automatically and automatically.

In this work, design flaws are categorized in two groups. The first one, the structural design flaw, is the structural characteristic of an entity software system that expresses a deviation from a given set of criteria typifying the high quality of a design such as Switch Statement and Refused Bequest. The second group, the behavioral design flaw, is the subjective characteristic which cannot be directly expressed in the structural appearance. Examples of this kind of flaw are Long Parameter List and Data Clumps.

2.2   Content Ontology

The next important term is *ontology*, which describes the terminology applied in a specific domain to represent the concepts contained therein [7, 9]. It can be read by machines, and encompasses the terminology of the domain and the relationships between concepts within that domain. The growth of the internet has led to increased sharing of knowledge, but this must be networked to achieve the greatest benefits. Ontologies simplify this process by shifting the emphasis away from the mere technological and towards a focus on knowledge in combination with technology. This is a vital distinction, and one which serves to make that knowledge far more practically beneficial in electronic formats.

Ontology can be shown as quintuplet: C (Concept), R (Relation), F (Function), A (Axiom) and I (Instance) composition, as $O = < C, R, F, A, I >$.

Concept (class) is the description and abstraction from different levels of domain knowledge, which covers all the attributes of the objective, it is the aggregate of all the instances that $C_i = \{I_{i1}, I_{i2}, I_{i3}, \ldots, I_{ij}\}$. Where I_{ij} is the instance contained in the concept.

Relation expresses the relationship between the concepts; it is used to describe the constitutive relations, and is presented by the subset of the N-dimensional Cartesian product, as R ($R \subseteq C_1 \times C_2 \times \ldots \times C_i$), that constructed the hierarchy between the concepts. Ontologies are defined as four categories based on semantic relationships: the whole-part relation (part-of); inheritance relation (kind-of); concept example relation (instance-of); property relation (attribute of).

Domain ontologies describe specific fields, while different fields each have their own ontologies. In terms of information and communication, knowledge is stored in a manner similar to a regular relational database, but the semantic organization is enhanced through logical construction and an emphasis on the relationships between entities, making the result superior to a normal database.

For the purpose of this study, the creation of an ontology for flaw-explicit descriptions requires the knowledge involved in the identification of flaws. The structure of this description knowledge comprises an inside level and an outside level. The former is the OWL ontology, or internal logic level, while the latter is constructed on the basis of the internal logic level and is composed of the Semantic Web Rule Language (SWRL) which serves the purpose of representing the identification knowledge which is used for reasoning.

3   The Proposed Approach of Flaw Detection

In this section, the proposed approach for design flaw detection is shown in Fig. 1. The approach shows eight steps of the proposed detection approach. The first four steps are in the Representation phase which is responsible for building an explicit description of flaw concepts and representing the desired logic programs of design flaw detection. The last four steps are in the Detection phase which is responsible for detecting design flaws in the object-oriented software. Step 1 and Step 5 are generic and must be based on a representation set of elements and relations of object-oriented concepts. Steps 2 to 4 must be followed when a new flaw is specified. Steps 5 to 8 are repeatable and must be applied on the source code of software systems.

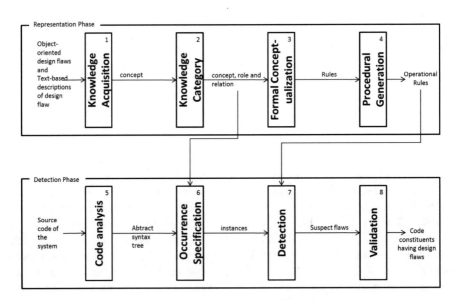

Fig. 1. The proposed approach of design flaw detection

The research considers ontology-based meaning from flaw domains that are perfect, that is, these domains are correct and complete. To show how to represent rules and to detect design flaws by such rules, the proposed approach performs the following step-by-step procedure. The proposed methodology is based on a declarative meta programming technique [5]. Each step is explained by clear presentation which is based on common patterns: input, output, and a description of each step.

3.1 Step 1: Knowledge Acquisition

Input:

1. Domains which are derived from design principles and heuristics [1, 3].
2. Examples of design flaw.
3. Target concepts of design flaws used to be described.

Output: Related ontology of flaws domains which are sufficient to describe and to generate detection rules of such flaws. All of information outputs are in the form of class ontology.

Description of step 1: The first step deals mainly with the arrangement of domain theories and a target concept to be represented for detection rules (Step 1 to Step 4). This step begin with each an example of flaw. Then, domain theories related to ontology components are defined. The research define such information class according to principles and heuristics of object-oriented paradigm. For the purpose in flaw detection, domain theories are any set of prior beliefs about the object-oriented design and implementation principles.

3.2 Step 2: Knowledge Category

Input: Ontology classes which relate to object-oriented flaws.

Output: Classes, slots, roles and relations for generating detection rules.

Description of step 2: Given the information in the first process, the second process is to determine a generalization of ontology concepts which is a sufficient concept definition for the flaw concept. The process provides the ontology building. Slots, roles and relations of all determined classes is represented in form of the ontology structure. To see more concretely how the approach works, consider representing concept, slots, roles and relations of Class domain in Fig. 2.

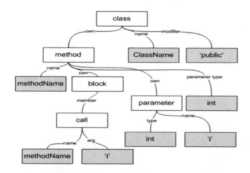

Fig. 2. Classes, slots, roles and relations of class domain

3.3   Step 3: Formal Conceptualization

Input: Classes, slots, roles and relations of the ontology structure.

Output: A logic detection rule from the ontology structure by the explanation.

Description of step 3: The ontology constructed by the previous step is generalized with a rule that is the most general relevant to the target concept (flaw concept). The research computes the most general rule that can be justified by the explanation in declarative meta programming, by computing the weakest preimage of the explanation. The weakest preimage of the target concept is computed by a general procedure called regression. The regression procedure operates on a domain theory represented by an arbitrary set of Horn clause. It works iteratively backward through the explanation, first computing the weakest preimage of the target concept with respect to the final proof step in the explanation, then computing the weakest of the resulting expressions with respect to the preceding step, and so on. The procedure terminates when it has iterated over all steps in the explanation.

3.4   Step 4: Procedural Generation

Input: A logic rule from the proof tree.

Output: New formulated rules.

Description of step 4: In this step, logic rules are refined by generalizing rules in accordance with a logic rule from the proof tree. Rules are pruned some literals for making generalization. At each learning cycle for generating a logic detection rule (Step 1–4), the sequential covering algorithm of ontology concepts picks a new positive representation that is not yet covered by the current rules and formulates a new rule according to the learning cycle. When the research provide more concepts, the rule of flaw is refined to be more general that the attribute of mutator method and accessor method cannot be the same attribute.

3.5   Step 5: Code Analysis

Input: Object-oriented source codes which need to detect design flaws.

Output: Representation tree which belong to AST specification.

Description of step 5: In this step, source code is parsed and formed in Abstract Syntax Tree (AST). It represents syntactical and semantical information of source code. The representation of source code is a tree of nodes which represents constants or variables (leave node) and operators or statements (inner nodes).

3.6   Step 6: Occurrence Specification

Input: AST logic facts.

Output: Logic facts which belong to defined ontology knowledge.

Description of step 6: These trees are transformed according to ontology-based fact structure. The argument in predicate calculus is represented by leave node and predicate part is represented by inner nodes of AST respectively. The number in each fact is used to represent relations among elements.

3.7   Step 7: Detection

Input: Logic facts and formulated logic rules.

Output: Results of the detection in each flaw.

Description of step 7: The detection of design flaw takes place in this step. The source code used for design flaw detection is transformed to first order logic facts. The detection performs by using pattern matching mechanism between facts and rules. Backward chaining mechanism in this environment performs searching to get the results. The detection algorithm halts once it finds the first valid proof.

3.8   Step 8: Validation

Input: Results of the detection.

Output: Precision and other rates of the detection by the proposed approach.

Description of step 8: Results of the proposed detection approach are validated by analyzing the suspicious types in the context of the complete approach of the system and its environment. The validation is inherently a manual task. Therefore, the detection of a few design flaws in different behavioral types is chosen to apply.

The research uses precision to show the number of true identified flaws, and false positive for the number of false positive missed by the detection. Two such rates are shown in the Eqs. (1) and (2).

$$\text{Precision} = \frac{true\ positive}{true\ positive + false\ positive} \tag{1}$$

$$\text{False positive} = \frac{false\ positive}{true\ positive + false\ positive} \tag{2}$$

4   Experiments and Discussions

In this section, this work starts by briefly describing the approach in the real detection by creating the prototype. Then, some case studies on various-sized programs are presented. Some interesting points from the proposed approach are also discussed.

4.1   Prototype

The work aim to validate the detection approach by specifying and detecting many varied types of flaws and computing the precision and some rates of the a generated prototype on CommonCLI v1.0 and JUNIT v1.3.6, two open source systems:

CommonCLI v1.0 is the Apache Commons CLI library. It provides an API for parsing command line options passed to programs. It contains 18 classes and 4132 lines of code. JUNIT v1.3.6 is a testing framework which performs automated testing. It contains 111 classes and 5,000 lines of code.

The prototype model is implemented for design flaw detection. The research use Eclipse v3.6 HERIOS, Prolog Development Tools v0.2.3 and SWI-Prolog v5.8.3 for implementing the prototype. All computations are performed on an Intel platform - i5 Intel at 2.44 GHz with 4 GB of RAM.

4.2   Results and Discussions

Figure 3a and b report numbers of true flaw and true detection of six flaws, respectively for the CommonCLI and JUNIT systems. Figure 4 shows precision rates and false positive rates of six flaws for such two software systems.

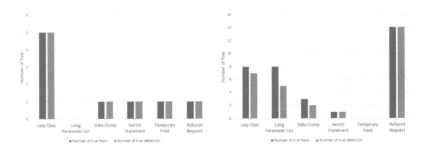

(a) The result of true flaws and true detection with six flaws in in CommonCLI v1.0

(b) The result of true flaws and true detection with six flaws in JUNIT v1.3.6

Fig. 3. The result of true flaws and true detection

After analyzing the resulting data, the work presents some important points to discuss:

For the result data from the proposed approach, the validation shows that the proposed approach of design flaw detection shows the expected precision and a good false positive, especially with structural design flaw detection.

In the details of Figs. 3a, b and 4, the proposed detection is performed on six types of design flaw. With Lazy Class and Data Clump the precision rate is 100%, with the exception of Long Parameter List which does not exist in CommonCLI. In addition, the last three structural design flaws - Switch Statement, Temporary Field, and Refuse Bequest all show a 100% precision rate. In the details of flaw detection for JUNIT, precision rates in behavioral design flaw detection are 87.50%, 62.50% and 66.66% for

Lazy Class, Long Parameter List, and Data Clump respectively. This confirms a 100% precision rate of all flaws for the last three structural design flaws. The results from the detection have, mainly, an average precision rate in detection of 100% with the flaw detection in CommonCLI and 85% with the flaw detection in JUNIT.

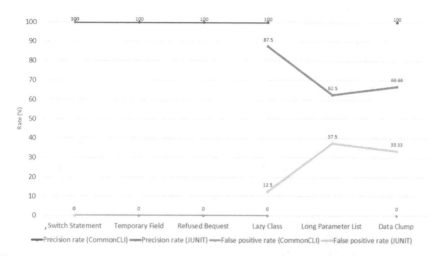

Fig. 4. Precision rates and false positive rates of six flaws in CommonCLI v1.0 and JUNIT v1.3.6

In terms of threats to the validity, the validity of the results depends directly on the flaw specifications of the rules. The research involved experiments on representative examples of flaws to lessen the threat to the internal validity of the validation. However, a 100% precision rate of the proposed methodology is difficult to obtain with complex behavioral design flaws because such flaws are typically subjective (although reasonable). The threat to external validity is related to exclusive use of open-source systems, which may limit the generalization of results to other systems. The subjective nature of interpreting and specifying flaws and related domain theory is a threat to the construct validity of this validation.

5   Conclusion and Future Works

In this paper, the research proposed a design flaw identification methodology by using ontology-based representation with a declarative meta programming technique. The explanation interprets examples of interesting design flaws, bringing into focus alternative coherent sets of features which might be important for correct classification. With this proposed detection, rules generated from ontology representation facts can detect flaws more conveniently than other approaches in the case of automatic detection and ignorance of proper threshold consideration and are effective in cases where the precision rate for the results is good – especially with structural design flaw detection – and it is not inferior to other approaches.

The validation in terms of precision and false positive sets some future positions. The work plans to perform such a comparison of this work with previous approaches in different design flaws. Further research will also plan to look for more domain theory information which can support the representing mechanism of ontology to increase the detection accuracy rate of behavioral design flaw detection. It will also implement tools and study the meta model of design flaw logic representation in terms of further specification of the detection model.

Acknowledgments. This research received funding from University of Phayao (Project No. RD60037) and was supported in part by the School of Information and Communication Technology, University of Phayao, Thailand.

References

1. Brown, W.J., Malveau, R.C., McCormick III, H.W., Mowbray, T.J.: AntiPatterns: Refactoring Software, Architectures, and Projects in Crisis. Wiley, New York (1998)
2. Coad, P., Yourdon, E.: Object-Oriented Analysis, 2nd edn. Yourdon Press, UpperSaddle River (1991)
3. Fowler, M., Beck, K., Brant, J., Opdyke, W., Roberts, D.: Refactoring: Improving the Design of Existing Code. Addison-Wesley Professional, Boston (1999)
4. Mantyla, M., Vanhanen, J., Lassenius, C.: A Taxonomy and an Initial Empirical Study of Bad Smells in Code, pp. 381–384 (2003)
5. Mens, T., Wuyts, R., De Volder, K., Mens, K.: Declarative meta programming to support software development: Workshop report. SIGSOFT Softw. Eng. Notes **28**(2), 1 (2003)
6. Moha, N., Gueheneuc, Y.G., Duchien, L., Le Meur, A.F.: Decor: a method for the specification and detection of code and design smells. IEEE Trans. Softw. Eng. **36**(1), 20–36 (2010)
7. Noy, N.F., McGuinness, D.L.: Ontology Development 101: A Guide to Creating Your First Ontology. Online (2001)
8. Riel, A.J.: Object-Oriented Design Heuristics, 1st edn. Addison-Wesley Longman Publishing Co. Inc, Boston (1996)
9. Stuckenschmidt, H., Klein, M.: Reasoning and change management in modular ontologies. Data Knowl. Eng. **63**(2), 200–223 (2007)
10. Webster, B.F.: Pitfalls of object-oriented development. M T (1995)
11. Wheeler, D.A., Brykczynski, B., Meeson Jr., R.N. (eds.): Software Inspection: An Industry Best Practice for Defect Detection and Removal, 1st edn. IEEE Computer Society Press, Los Alamitos (1996)

Virtual Reality Application for Animal Cruelty Education

Napatt Kuttikun[(✉)], Pichayaporn Choosakulchart,
and Narit Hnoohom

Image, Information and Intelligence Laboratory, Department of Computer
Engineering, Faculty of Engineering, Mahidol University, Nakorn Pathom,
Thailand
napatt.kuttikun@gmail.com, pichaya_om@hotmail.com,
narit.hno@mahidol.ac.th

Abstract. Educational virtual reality applications were developed to highlight
animal cruelty in the livestock production industry using Unity 3D Engine with
C# scripts and HTC Vive headsets. Users were subjected to pre and post virtual
reality assessments to measure their knowledge concerning the topic and their
feelings regarding the application. They scored an average of 22.54% for pre-
assessment and 65.62% for post-assessment; 86.7% felt better informed after the
application, 70% felt entertained, and nobody was bored. Future improvements
regarding application components and expansion of applicable topics are
planned.

Keywords: Virtual reality · HTC Vive headset · Animal cruelty education
Unity 3D Engine

1 Introduction

Millions of animals are injured, tortured, and slaughtered daily in the animal production
industry. They are jostled, moved violently into small spaces and forced to live in
crowded conditions [1]. Education is key to allow people to make informed decisions
regarding animal cruelty. Improved public interaction will increase engagement and
emotional attentiveness. Stimulating perception using an artificial environment will
also increase emotional connections with animals [2]. Virtual reality (VR) has the
capacity to accomplish these improvements. This paper developed a virtual reality
application to educate the public regarding animal cruelty in the livestock production
industry. An artificial environment was created using 3D software with a HTC Vive
headset and an HTC Vive controller for user interaction.

Virtual reality has been used before as an educational medium. Fowler [1]
explained in the British Journal of Educational Technology that virtual reality offers
immersion in terms of conception, task, and social. These three types of immersion lead
to an improved learning experience. A Tech trends journal article published by Brown
and Green [3] noted that virtual reality has been successfully used in training simu-
lations, virtual explorations, and for raising cultural values. VR has the ability to
change user thoughts and actions. Low cost hardware as virtual reality headsets is

© Springer Nature Switzerland AG 2019
T. Theeramunkong et al. (Eds.): iSAI-NLP 2017, AISC 807, pp. 129–142, 2019.
https://doi.org/10.1007/978-3-319-94703-7_13

widely available including GearVR and Google Cardboard. Unity 3D is also a recommended free software for developing virtual reality applications. Heradio et al. [4]
agreed in the Journal of Computer & Science that virtual reality is a low cost tool to
simulate laboratory work in engineering and scientific education. VR is also easy to
maintain and safer to operate compared to conventional methods. Moreover, Psotka
[5], a professor at the Army Research Institute for the Behavioral and Social Sciences,
confirmed that virtual reality brings motivational learning to students, even after a long
period of usage. VR also has the ability to transform intangible and abstract concepts to
tangible experiences. Creating concrete experiences in a virtual world can help students
to better understand the concepts presented.

The hardware and software selection for the application are discussed followed by
the development methods used. Results are discussed and conclusions drawn in the
final section.

2   System

This application comprised two main components of hardware and software. In
hardware part, we used the HTC Vive which is a greatest suite of virtual reality
hardware development tools. The related software of the paper is shown in Fig. 1,
contains two programs: Blender, Unity and Microsoft Visual studio.

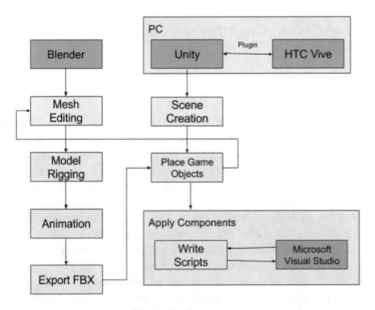

Fig. 1. System process

2.1 Hardware Selection

Figure 2 shows the overall hardware system architecture. All components of the HTC Vive (headset, base stations, and controllers) require connection to a personal computer (PC). The minimum requirements for this PC are shown in Table 1 [6].

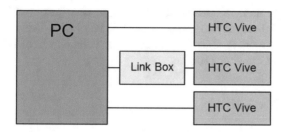

Fig. 2. System architecture

Table 1. Minimum requirements for HTC Vive

Item	Hardware requirements
Processor	Intel™ Core™ i5-4590 or AMD FX™ 8350, equivalent or better
Graphics	NVIDIA GeForce™ G TX 1060 or AMD Radeon™ RX 480, equivalent or better
Memory	4 GB RAM or more
Video output	1x HDMI 1.4 port, or DisplayPort 1.2 or newer
USB	1x USB 2.0 port or newer
Operating system	Windows™ 7 SP1, Windows™ 8.1 or later or Windows™ 10

When connecting each component to the PC, certain requirements must be met. Each base station must be within each other's range. If this is not possible, a cable can be connected to sync the stations. The Link Box must be supplied with power from a power cord, and connected to the PC via HDMI and USB cables. A 3-in-1 tether is provided to attach the headset to the Link Box. The entire process of connecting the headset, Link Box, and PC is shown in Fig. 3.

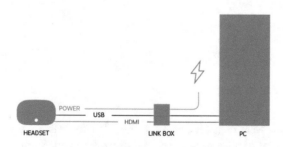

Fig. 3. Connecting the HTC Vive headset to a PC

Steam VR is required to synchronize all the components of the HTC Vive to each other [7]. Following the instructions for the basic setup allows all the components to be connected, either wired or wirelessly.

2.2 Software Selection

The main software used to develop the application was Unity 5.5.3 since it is compatible with HTC Vive development. SteamVR Plugin for Unity was also another major software component to access the hardware components of HTC Vive such as controllers, headset tracking and play area.

3 Proposed Methods

Each component was chosen to maximize the learning experience for the users. Models were chosen mainly as decorations, making the overall experience less mundane. The application flow and map layout were designed to be self-explanatory, making everything easier for users to understand. A written code was applied to many parts of the interactions throughout the application, offering knowledge as well as entertainment.

3.1 Model Selection

Model selection for this application comprised three main components as NonPlayable-Characters, Animals and Other models.

Non-playable-characters. Cartoon Assets published by Synty Studios, shown in Fig. 4. Character avatars were incorporated with the character controller and scripts from Unity along with dialogues for speech. The character avatars interacted with the players throughout the application and guided them through events.

Fig. 4. Simple farm: cartoon assets by Synty Studios [8]

Animals. Animals were controlled using scripts. The models came with animations to make them more realistic. Cartoon Assets published by Synty Studios, shown in Fig. 5.

Fig. 5. Simple farm animals: cartoon assets by Synty Studios [9]

Other Models. Environmental assets that complemented the overall experience of the user were added throughout the application, as well as tools that players used to complete tasks.

3.2   Map Design

The application setting was a farm with an initial tutorial to get used to the controls. Once the player progressed through the application, three different factories appeared housing three different animals to learn about.

Chicken

- Baby chicks: Users learned about the living conditions of chicks and what they experience after birth.
- Chicken growth: Users learned about chickens and their living conditions, as shown in Fig. 6.
- Chicken living area: Users learned about how the chickens were slaughtered.

Fig. 6. Chicken growth scene

Pigs

- Piglets: Users learned about the birthing conditions of the mother and living conditions of the piglets.
- Pig growth: Users learned about the living conditions of pigs raised for meat, as shown in Fig. 7.
- Pig slit area: Users learned about the slaughter process for pigs, as shown in Fig. 8.

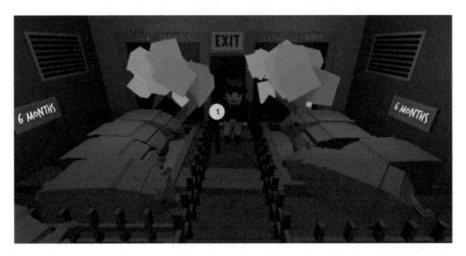

Fig. 7. Baby piglets follow their mother as she is being dragged out

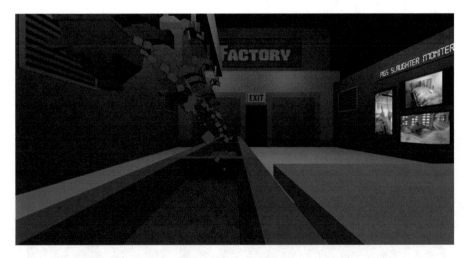

Fig. 8. Baby piglets follow their mother as she is being dragged out

Cows

- Dairy cows: Users learned about the different stages of the dairy cycle for cows and their living conditions, as shown in Fig. 9.
- Baby cows: Users learned about how and why veal meat was produced and how the animals were slaughtered.
- Cow growth: Users learned about the living conditions of cows raised for meat.

Fig. 9. Dairy cow scene

3.3 Pseudocode

Scripts control various aspects of the application including animation, lights, scene management, interactions, and physics. The pseudocode of the main scripts used in the application was as follows:

The first pseudocode was used to control the interaction of the user with HTC Vive, as shown in Algorithm 1. This code was attached to both of the controllers in all scenes.

Algorithm 1. Pseudocode for HTC Vive controller

SET controllers device

<u>FUNCTION</u> OnTriggerStay()

  //called when object's collider trigger controller's collider

IF controller is not holding anything

IF controller's grip pad is pressed

IF colliding object has tag 'pickup'

SET colliding object's parent to controller's parent

SET controller has picked up something

ENDIF

    ELSE controller's grip pad is released

IF controller is holding something

 SET colliding object's parent to null

 ENDIF

    ENDIF
ENDIF

The second pseudocode was used to control the dialog box that appeared on the screen of the application as shown in Algorithm 2. Dialog progressed when the user pressed the trigger on the controller and 5 s passed until it appeared. This script was attached to the user's headset in all scenes.

Algorithm 2. Pseudocode for dialog controls

INITIALIZE textbox attached to HTC Vive headset
SET iTimeLeft to 5 seconds
SET iCurrentLine to 1
FUNCTION Update()

   //called every frame when the application is running
SET textbox = DialogProgress(iCurrentLine)
SET reduce iTimeLeft
IF iTimeleft is 0 and wait event trigger is false
SET iTimeLeft to 5 seconds
SET increase iCurrentLine by 1
ENDIF END FUNCTION

FUNCTION DialogProdress(iCurrentLine)
SWITCH(iCurrentLine)
CASE 1: Dialog for first line
CASE 2: Dialog for second line

SET Wait for trigger event true

       Wait for user's interaction to set wait for trigger event to false

….

CASE 15: Last dialog for the scene
SET scene manager to load next scene

The third pseudocode was used to control animal movement to run away from the users. This script was used in the chicken and pig scenes and attached to the animal objects that ran away from the direction of the user's headset.

Algorithm 3. Pseudocode to control animal movement

    SET runaway to false
    SET speed

    <u>FUNCTION</u> OnTriggerStay **IF**

    animal runaway is true

    IF animal collide with user's headset
    SET running animation for animal

    SET moveDirection to be animal's direction subtracted by user's direction
    SET rotate animal to be moveDirection in the X and Z axis only

        //no vertical rotation in the Y axis
   SET movement of animal to moveDirection times by speed

---

4 Experiment Results

A total of 30 users participated in the experiment. Users were subjected to a pre-assessment that asked several questions about each of the animals, including comparisons between natural and genetically modified animals, their living conditions, and other information. After the pre-assessment, players used the application and then answered the same set of questions in the post-assessment. Both scores were averaged and compared. Other information regarding users' opinions on the application and their overall experience was collected after the post-assessment.

4.1 System Results

The application began with a tutorial informing the players about the controls through doing simple farming tasks. The players were given three options to select the story of the animal that they wanted to follow as shown in Fig. 10. Players then completed the events within each animal's route, learning about the animals' birth, living conditions, slaughter, and other information. Once a route was completed, the player selected another route to follow.

4.2 Experiment Results

The 30 users tested recorded an average score of 22.54% during the pre-assessment but 65.62% after the post-assessment at over 2 times higher, as shown in Table 2.

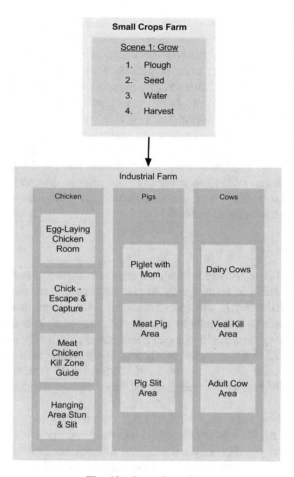

Fig. 10. Story flow chart

The median also increased from 2 points to 9 points, showing a 4.5 fold increase. This indicated that the application was successful in increasing the overall knowledge level of the users. Thus, similar applications can be created in the future involving other subjects to produce similar positive results.

Table 2. Statistical summary of the pre-and post-assessment

	Pre-assessment	Post-assessment
Average	2.93/13 (22.54%)	8.53/13 (65.62%)
Median	2/13 (15.38%)	9/13 (69.23%)
Range	0–8 (8)	4–13 (9)

The range remained almost constant, being 8 in the pre-assessment and 9 in the post-assessment. However, minimum and maximum scores of both increased. The minimum score of 0% during the pre-assessment increased to 30.77% by the post-assessment, and the maximum score of 61.54% on the pre-assessment increased to 100% on the post-assessment. However, certain individuals still gained very low scores during the post-assessment, while some scored much higher than the average. The range and score spread were substantial. This result was similar to a traditional bell curve, showing that the application was still highly affected by the user. Users with higher levels of interest and greater motivation to achieve a high score were more likely to succeed. Since the application relayed information through the English language, users with lower levels of English proficiency encountered difficulty achieving a high score. This indicated that, as in a real classroom, language proficiency was very important for the education of users.

4.3 Discussions

Most users at 66.8% tested had no prior experience with virtual reality; however, the overall difficulty rating in learning how to play averaged. Thus, despite being a new technology, virtual reality is not difficult to adopt. With a well-built tutorial system, users should be able to learn how to play quickly, with a rapid understanding of the overall virtual reality system. On average, female users gave a difficulty rating of 2.24, while males scored 2.29, showing no significant difference between genders. This indicated that gender played no role in the difficulty of system adoption which depended on individual user experience and other factors requiring more in-depth research.

When asked how they felt, 86.7% of users reported that they gained knowledge from the application, 70% were entertained and none were bored. This indicated that the application had potential as a learning tool that was still somewhat entertaining. Additionally, 43.3% also felt emotional, showing that under a more emotive context, virtual reality has the potential for use in topics requiring emotion; 33.3% of users also reported feeling tired afterward, so the application may not be fit for those who cannot withstand light exercise for extended periods of time. Despite the learning effect of the application being heavily reliant on user attention and motivation level, there is a possibility that because of the entertaining nature, user attention and motivation might be higher compared to other learning experiences, thus causing them to perform better when learning through the application. However, more research is required on this subject for a clear conclusion. While some users did not score highly on the post-exam, 86.7% still reported feeling informed, indicating a positive gain in general knowledge, rather than specific information asked for during the assessments.

The average user immersive rating was 3.87 out of 5, showing that the application had high immersive properties. This proved that the low-quality cartoon-like graphics of the application did not hinder the overall immersive user experience. More research is required, specifically regarding the comparison between cartoon-like and hyperrealistic graphics within a virtual reality setting.

From the valid responses (excluding vegetarians and vegans), 44.44% said they would consider reducing their meat consumption, while none considered going vegetarian or vegan. This showed that while the application may be able to raise awareness, it was not effective in causing drastic changes to the primary diet of users.

5   Conclusions

This paper discussed the system and method used to develop an educational virtual reality application concerning animal cruelty in the livestock production industry. Pre- and post-assessment scores were measured as well as user emotions after experiencing the application. Data regarding the immersiveness rating, gender, and change in diet of the users were also collected.

Results indicated that this application has the potential to educate users about the topic while keeping them somewhat entertained and emotional. However, the degree of improvement still depends on factors such as user level of interest. Most users felt entertained and informed while none reported feeling bored. The application, therefore, shows potential as a learning tool that remains somewhat entertaining and can educate and raise awareness without impacting on dietary change. Obstacles that users experienced included getting comfortable with the virtual reality environment and learning how to control it. Language was another main barrier for the users.

In the future, this application can be improved by expanding into other virtual reality platforms to include other languages. The models and map design can also be upgraded to increase immersiveness. Similar techniques can be applied to different types of applications, both educational and non-educational. Virtual reality applications can be used in medical practices, military practice, and educational practice.

Acknowledgment. This work was supported by the Department of Computer Engineering, Faculty of Engineering, Mahidol University.

References

1. Fowler, C.: Virtual reality and learning: where is the pedagogy? Br. J. Educ. Technol. **46**(2), 412–422 (2015). https://doi.org/10.1111/bjet.12135
2. Julia, D., Georg, W.A., Henrik, M.P., Youssef, S., Andreas, M.: The impact of perception and presence on emotional reactions: a review of research in virtual reality. Front. Psychol. **6**, 26 (2015). https://doi.org/10.3389/fpsyg.2015.00026
3. Brown, A., Green, T.: Virtual reality: low-cost tools and resources for the classroom. Techtrends Link. Res. Pract. Improv. Learn. **60**(5), 517–519 (2016). https://doi.org/10.1007/s11528-016-0102-z
4. Heradio, R., de la Torre, L., Galan, D., Cabrerizo, F.J., Herrera-Viedma, E., Dormido, S.: Virtual and remote labs in education: a bibliometric analysis. Comput. Educ. **98,** 14–38 (2016). https://doi.org/10.1016/j.compedu.2016.03.010
5. Psotka, Joseph: Educational games and virtual reality as disruptive technologies. Educ. Technol. Soc. **16**(2), 69–80 (2013)

6. VIVE™|Vive Ready Computers, Vive.com (2016). http://www.vive.com/us/ready/. Accessed 20 Nov 2016
7. HTC Vive PRE Installation Guide - SteamVR - Knowledge Base - Steam Support, 1Support. steampowered.com   (2016).   https://support.steampowered.com/kb_article.php?ref=2001-UXCM-4439. Accessed 20 Nov 2016
8. Asset Store, Unity Asset Store (2017). https://www.assetstore.unity3d.com/en/#!/content/47003. Accessed 29 Mar 2017
9. Asset Store, Unity Asset Store (2017). https://www.assetstore.unity3d.com/en/#!/content/76405. Accessed 29 Mar 2017

Business Intelligence System

Anti-theft Motorcycle System Using Face Recognition by Deep Learning Under Concept on Internet of Things

Apichat Silsanpisut$^{(\boxtimes)}$, Patcharaon Petchsamutr, and Mahasak Ketcham

Faculty of Information Technology, King Mongkut's University of Technology North Bangkok,
Bangkok, Thailand
{s5907021857147,s5807021857521}@email.kmutnb.ac.th,
maoquee@hotmail.com

Abstract. The special problem aimed at to develop an anti-theft motorcycle system using face recognition by deep learning under concept on internet of things for use in the prevention of theft motorcycle. The special problem is used this technique face detection, face recognition and internet of things by measuring the image received from the camera to compare the original owners of the motorcycle when it detects that the original image and the image received from the camera that does not match.

The results of system are equipment and incoming image from the camera accurately for increase the effectiveness of the theft motorcycle and the system can used in current.

Keywords: Motorcycle · Face recognition · Deep learning · Internet of Things

1 Introduction

A motorcycle statistics for motorcycles registered in Thailand with the Department of Land Transport at 31 December 2015 [1] total of 35,546,520 and motorcycles for a person of 20,308,201 units and motorcycle public sign of 189,362 units as well as two types of 20,407,563 units of the motorcycle is as a vehicle is due to the cars are the traffic jamthe motorcycle to be able to help people achieve the best vehicles is small, will be able to use shortcuts to get to it at the appointed time and the widely used in all, a career path.

Statistics from the Royal Thai Police on the grand theft auto, motorcycle information in [3] 1020-1002 identifies the number of that the theft more than but the return of that (Fig. 1).

© Springer Nature Switzerland AG 2019
T. Theeramunkong et al. (Eds.): iSAI-NLP 2017, AISC 807, pp. 145–158, 2019.
https://doi.org/10.1007/978-3-319-94703-7_14

Fig. 1. To pry open the lock, a motorcycle (left) and the direct line of motorcycles (right).

Sreedevi and Sarath (2011) The system described in this paper automatically take photos of driver and compares his or her face with database to check whether he is an authenticated driver or not. He can have access to the vehicle only if he is an authenticated driver. If he is not an authenticated driver an alarm rings and electrical connections are not activated. The technology used here is face recognition and face detection in real time. As the photos are taken in real time, several problems like unequal illumination and changes in the background may affect the system. To overcome this problem DCT normalization and background cancellation algorithms are incorporated along with basic face detection and face recognition algorithm. In this paper software and hardware details of the system are discussed. Technologically system is simple, accurate and maintainable.

Bagavathy, Dhaya and Devakumar (2011) The existing system was. Car alarm techniques are used to prevent the car theft with the help of different type of sensors like pressure, tilt and shock & door sensors. Drawbacks are cost and cant used to find out the thief, it just prevents the vehicles from loss. The proposed security system for smart cars used to prevent them from loss or theft using Advanced RISC Machine (ARM) processor. It performs the real time user authentication (driver, who starts the car engine) using face recognition, using the Principle Component Analysis - Linear Discriminant Analysis (PCA LDA) algorithm. According to the comparison result (authentic or not), ARM processor triggers certain actions. If the result is not authentic means ARM produces the signal to block the car access (i.e. Produce the interrupt signal to car engine to stop its action) and inform the car owner about the unauthorized access via Multimedia Message Services (MMS) with the help of GSM modem. Also it can be extends to send the current location of the vehicle using the GPS modem as a Short Message Services (SMS) as passive method [10].

The researchers need to have to uses pattern face recognition by deep learning with the loss in a motorcycle to prevent the loss, parked in a secluded parking lot in the parking lot due to recognition in the face can also determine the identity of the robbery. By this research to develop robbery prevention by face recognition through the work of JetsonTX2 Develop kit of online applications to the owner of the motorcycle.

2 Proposed Method

We've used machine learning to solve isolated problems that have only one step estimating the price of a house, generating new data based on existing data and telling if an image contains a certain object. All of those problems can be solved by choosing one machine learning algorithm, feeding in data, and getting the result. But face recognition is really a series of several related problems (Fig. 2).

1. First, look at a picture and find all the faces in it.
2. Second, focus on each face and be able to understand that even if a face is turned in a weird direction or in bad lighting, it is still the same person.
3. Third, be able to pick out unique features of the face that you can use to tell it apart from other people like how big the eyes are, how long the face is, etc.
4. Finally, compare the unique features of that face to all the people you already know to determine the person's name.

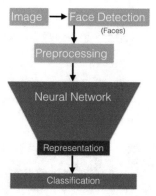

Fig. 2. The process of face detection with deep learning

As a human, your brain is wired to do all of this automatically and instantly. In fact, humans are too good at recognizing faces and end up seeing faces in everyday objects.

Computers are not capable of this kind of high-level generalization so we have to teach them how to do each step in this process separately.

We need to build a pipeline where we solve each step of face recognition separately and pass the result of the current step to the next step. In other words, we will chain together several machine learning algorithms (Fig. 3).

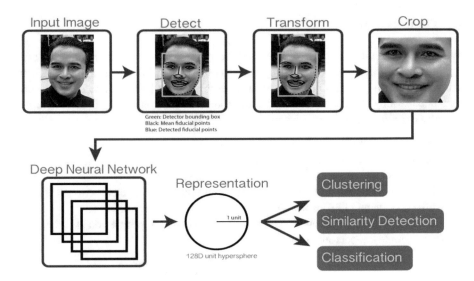

Fig. 3. The process of preprocessing into detect, transform and crop

2.1   Finding All the Faces

The first step in our pipeline is face detection. Obviously we need to locate the faces in a photograph before we can try to tell them apart (Fig. 4).

Fig. 4. Finding all the faces

Face detection is a great feature for cameras. When the camera can automatically pick out faces, it can make sure that all the faces are in focus before it takes the picture. But we'll use it for a different purpose finding the areas of the image we want to pass on to the next step in our pipeline.

We're going to use a method called Histogram of Oriented Gradients.

To find faces in an image, we'll start by making our image black and white because we don't need color data to find faces (Fig. 5).

Fig. 5. Find faces in an image, it's made our image black and white

Can be obtained from equation

$$G' = 0.3R + 0.59G + 0.11B \tag{1}$$

Define G' to the gray value
R to the green value
G to the green value
B to the blue value

Then every pixel in the image is displayed. For every pixel you want to see the pixels that surround you directly (Fig. 6).

Fig. 6. This steps to see the pixels in the image.

Our goal is to figure out how dark the current pixel is compared to the pixels directly surrounding it. Then we want to draw an arrow showing in which direction the image is getting darker (Fig. 7).

If you repeat that process for every single pixel in the image, you end up with every pixel being replaced by an arrow. These arrows are called gradients and they show the flow from light to dark across the entire image (Fig. 8).

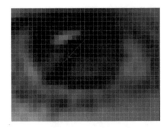

Fig. 7. The process is looking at just this one pixel and the pixels touching it, the image is getting darker towards the upper right.

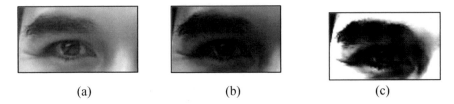

| (a) | (b) | (c) |

Fig. 8. Gradient in Pixel Image (a) Gradient Image (b) Gradient to Dark Spot (c) The gradient of light reaches the dark of all images.

For replacing the pixels with gradients. If we analyze pixels directly, really dark images and really light images of the same person will have totally different pixel values. But by only considering the direction that brightness changes, both really dark images and really bright images will end up with the same exact representation.

That makes the problem a lot easier to solve. But saving the gradient for every single pixel gives us way too much detail. We end up missing the forest for the trees. It would be better if we could just see the basic flow of lightness/darkness at a higher level so we could see the basic pattern of the image.

To do this, we'll break up the image into small squares of 16 × 16 pixels each. In each square, we'll count up how many gradients point in each major direction. Then we'll replace that square in the image with the arrow directions that were the strongest (Fig. 9).

Fig. 9. The original image is turned into a HOG representation that captures the major features of the image regardless of image brightness

To find faces in this HOG image, all we have to do is find the part of our image that looks the most similar to a known HOG pattern that was extracted from a bunch of other training faces (Figs. 10 and 11).

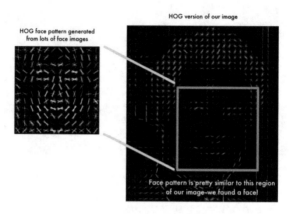

Fig. 10. Histogram of Oriented Gradients Algorithm

Fig. 11. The result of Histogram of Oriented Gradients Algorithm

2.2   Posing and Projecting Faces

The second step isolated the faces in our image. But now we have to deal with the problem that faces turned different directions look totally different to a computer (Fig. 12).

This Step will try to warp each picture so that the eyes and lips are always in the sample place in the image. This will make it a lot easier for us to compare faces in the next steps.

To do this, we are going to use an algorithm called face landmark estimation. There are lots of ways to do this.

68 specific points is called landmarks that exist on every face the top of the chin, the outside edge of each eye, the inner edge of each eyebrow, etc. Then we will train a

Fig. 12. Faces of individuals facing different directions.

machine learning algorithm to be able to find these 68 specific points on any face (Fig. 13).

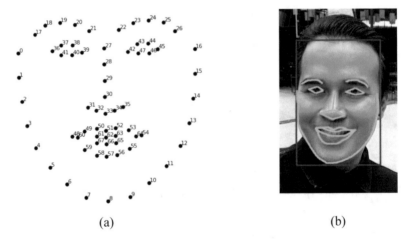

(a)   (b)

Fig. 13. (a) 68 specific point (b) the result of 68 specific point

Now that we know were the eyes and mouth are, we'll simply rotate, scale and shear the image so that the eyes and mouth are centered as best as possible. We won't do any fancy 3d warps because that would introduce distortions into the image. We are only going to use basic image transformations like rotation and scale that preserve parallel lines it's called affine transformations (Fig. 14).

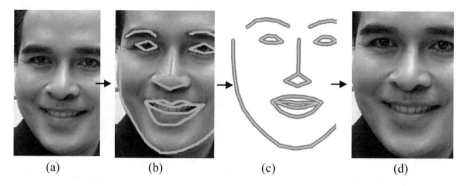

(a)   (b)   (c)   (d)

Fig. 14. (a) Face area in image (b) main face detect (c) the result of faces position required (d) the result closing of faces

Now no matter how the face is turned, we are able to center the eyes and mouth are in roughly the same position in the image. This will make our next step a lot more accurate.

2.3   Encoding Faces

This step separates faces from each other. The simplest approach to face recognition is to directly compare the unknown face we found in Step 2 with all the pictures we have of people that have already been tagged. When we find a previously tagged face that looks very similar to our unknown face, it must be the same person.

The Most Reliable Way to Measure a Face
Collecting face data to create our known database, ear size, nose length, eye color and other. The most accurate way is to make the computer calculate the data collection. Find the parts of the face that are important to measure.

The solution is to train a Deep Convolutional Neural Network. But instead of training the network to recognize pictures objects like we did last time, we are going to train it to generate 128 measurements for each face.

1. Load a training face image of a known person.
2. Load another picture of the same known person.
3. Load a picture of a totally different person.

Then the algorithm looks at the measurements it is currently generating for each of those three images. It then tweaks the neural network slightly so that it makes sure the measurements it generates for step 1 and step 2 are slightly closer while making sure the measurements for step 2 and step 3 are slightly further apart (Fig. 15).

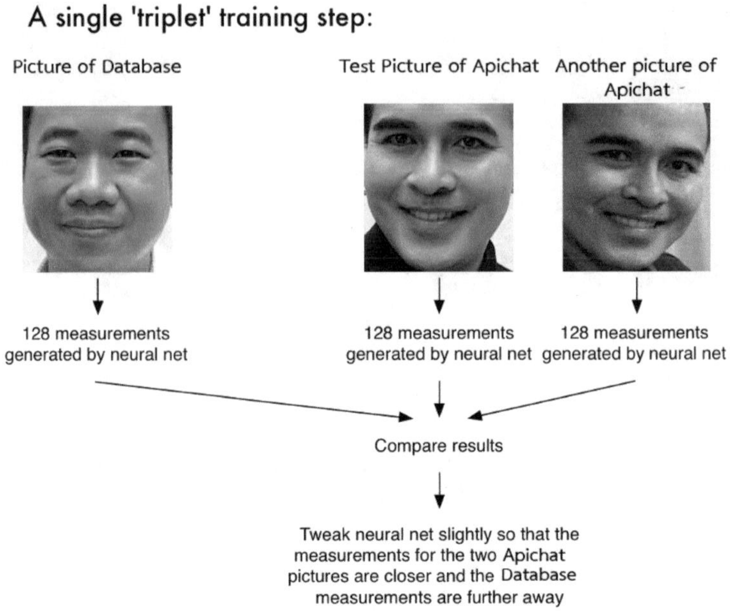

A single 'triplet' training step:

Picture of Database   Test Picture of Apichat   Another picture of
                                                                        Apichat

↓   ↓   ↓

128 measurements   128 measurements   128 measurements
generated by neural net   generated by neural net generated by neural net

Compare results

↓

Tweak neural net slightly so that the
measurements for the two Apichat
pictures are closer and the Database
measurements are further away

Fig. 15. The process of encoding face to training

After repeating this step millions of times for millions of images of thousands of different people, the neural network learns to reliably generate 128 measurements for each person. Any ten different pictures of the same person should give roughly the same measurements.

After repeating this step millions of times for millions of images of thousands of different people, the neural network learns to reliably generate 128 measurements for each person. Machine learning people call the 128 measurements of each face an embedding. The idea of reducing complicated raw data like a picture into a list of computer-generated numbers comes up a lot in machine learning.

Face Encoding

This process of training a convolutional neural network to output face embeddings requires a lot of data and computer power. Even with an expensive NVidia Telsa video card, it takes about 24 h of continuous training to get good accuracy. But once the network has been trained, it can generate measurements for any face, even ones it has never seen before! So this step only needs to be done once. So all we need to do ourselves is run our face images through their pre-trained network to get the 128 measurements for each face as shown Fig. 16.

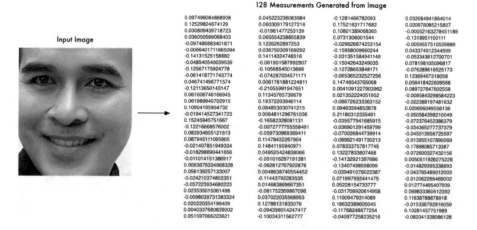

Fig. 16. The process of encoding for 128 measurements

2.4 Finding the Person's Name from the Encoding (Representation → Clustering/Classification/Detection)

This last step is actually the easiest step in the whole process. All we have to do is find the person in our database of known people who has the closest measurements to our test image. You can do that by using any basic machine learning classification algorithm. No fancy deep learning tricks are needed. We'll use a simple linear SVM classifier, but lots of classification algorithms could work. All we need to do is train a classifier that can take in the measurements from a new test image and tells which known person is the closest match. Running this classifier takes milliseconds. The result of the classifier is the name of the person (Fig. 17).

Apichat.png   **Apichat2.png**   **Apichat3.png**

Fig. 17. The result of the classifier is the name of the person.

2.5   Result

Rendering of processing results the system will display the results on the screen of the smartphone on the Android operating system and monitor display. The system is rendered by detecting a picture of the person who committed the motorcycle theft (Figs. 18 and 19).

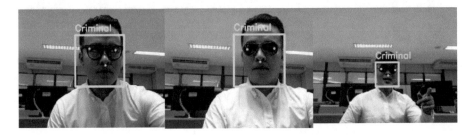

Fig. 18. The result of face recognition from decoding with deep learning

Fig. 19. The result of face recognition from decoding with deep learning on smartphone

2.6   Hardware

1. Jetson TX2 Developkit
    See Fig. 20

Fig. 20. Jetson TX2 Developer Kit

2. Smartphone
   See Fig. 21

Fig. 21. Smartphone

3   Result

The results of an Anti-theft motorcycle system using face recognition by deep learning under concept on internet of things. From testing system 20 times group by start system of Image Input, face detection, face recognition by deep learning as shown in Tables 1.

- From testing system for face detection of 20 times. The system can detect your face of 15 times and cannot detect your face of 5 times.
- From testing system for face recognition by deep learning of 20 times. The system can compare image between image input from camera and image of 18 times and cannot compare image of 2 times.

Tables 1. Anti-theft motorcycle system using face recognition by deep learning under concept on Internet of Things

Testing	Result
1. Face detection	75%
2. Face recognition by deep learning (purpose method)	90%

4 Conclusion

After the developed of Anti-theft motorcycle system using face recognition by deep learning under concept on internet of things was complete. The researcher conclude information of system.

- The results of system for face detection as 75%.
- The results of system for face recognition by deep learning. The system of Input image from camera is compared image. The owner of motorcycle or don't intend to be 90%.

However the system also found face detection, face recognition.

References

1. Motorcycle registration statistics, Department of Land Transport. https://data.go.th/DatasetDetail.aspx?id=21d8d6-378dfd-4a-12a26a-0dd1db308e24
2. Robbery, Office of the Royal Society. http://www.royin.go.th/?knowledges=โจรกรรม-๕-เมษายน-๒๕๕๓
3. Statistics motorcycle theft, Little Lee. http://www.oknation.net/blog/LittleLee/2011/03/07/entry-1
4. Guo, H., et al.: An automotive security system for anti-theft (2009)
5. Hameed, S.A., et al.: Car monitoring alerting and tracking model (2010)
6. Sirisak, L.: The Format of the Memory Card by Using Image Processing and Neural Network. Master of Engineering Thesis Major McKee electronic, Suranaree University of Technology, Thailand (2012)
7. Bagavathy, P., et al.: Real Time Car Theft Decline System Using arm Processor (2011)
8. Liu, Z., et al.: Vehicle Anti-theft Tracking System Based on Internet of Things (2013)
9. Amos, B., Ludwiczuk, B., Satyanarayanan, M.: OpenFace: a generalpurpose face recognition library with mobile applications, June 2016. CMU-CS-16-118
10. Bagavathy, P., Dhaya, R., Devakumar, T.: Real time car theft decline system using ARM processor. In: 3rd International Conference on Advances in Recent Technologies in Communication and Computing (ARTCom 2011) (2011)
11. Liu, S.-S., Tian, Y.-T., Li, D.: New research advances of facial expression recognition. In: 2009 International Conference on Machine Learning and Cybernetics, vol. 2, pp. 1150–1155 (2009)
12. Zhao, W., Chellappa, R., Phillips, P.J., Rosenfeld, A.: Face recognition: a literature survey. ACM Comput. Surv. 35, 399–458 (2003)
13. Lee, C., Landgrebe, D.: Feature Extraction and Classification Algorithms for High Dimensional Data. Purdue University (1993)

Balanced Scorecard Quality Information Dashboards Model for Competitive Business Advantage

Wiwit Suksangaram[1,2(✉)], Kritchana Wongrat[1,2],
and Sopaporn Klamsakul[1,2]

[1] Department of Computer Business, Phetchaburi Rajabhat University,
Phetchaburi, Thailand
Wiwit.suk@mail.pbru.ac.th, puwadolwon@yahoo.com,
Sopaporn_s@yahoo.com
[2] Department of Business Administration, Phetchaburi Rajabhat University,
Phetchaburi, Thailand

Abstract. The results showed that: (1) information Technology alliance of balanced scorecard to make a difference in the business. A total of 66 to share the information for senior executives of eight of the information technology executive with a total of 20 patents and information for management at the number 38 and by the many more who do not accept the information as significant. Statistic of 0.05.

All of this evidence. The innovative concept is considered in the analysis and design information that is appropriate for the agency. Creating a competitive advantage by using IT as a business tool. Using IT queue. To solve business problems, including quality tools. Air Quality Methodology Balance Scorecard. The integration of information science. Including analysis and design of information systems for organizations. And computer tools, including design input – processing –released. Correlation analysis of the information contained in each (stepwise Refinement) by the model information. That represents the relationship in the form of tables. Three-dimensional matrix, the people who want to bring this concept to the development of Information systems should understand before applying to the agency's next.

Keywords: Balanced scorecard · Quality information technology
Dashboards

1 Introduction

With current world changes, business operations have changed likewise. In the past, production of goods and services was intended to meet the basic needs, which has been transformed into industrial production [1], causing free, borderless competition. This results in a lot of competition, including competition between rivals in competing industries from abroad, new competitors entering the market because the present era is characterized by trade liberalization and borderless competition. As a result, the operators need to adapt to business changes. Business units need to develop the

© Springer Nature Switzerland AG 2019
T. Theeramunkong et al. (Eds.): iSAI-NLP 2017, AISC 807, pp. 159–168, 2019.
https://doi.org/10.1007/978-3-319-94703-7_15

organizations so as to be competitive with rivals in the industries with the ability to enhance the organizations' competitiveness, the capacity to enhance the performance, the ability to create quality goods and services, the ability to reduce operating costs, the ability to create customer satisfaction, the ability to develop work processes within the organizations, the ability to develop the potential of personnel, etc. This research is interested in the development of Balanced Scorecard Quality Information model for competitive advantage in the automotive industry on the issues of: (1) the importance of competitive advantage; (2) the problem of competitive advantage in the business sector; (3) the Balanced Scorecard quality methodology and the development of competitive advantage in business; (4) quality information technology; (5) Balanced Scorecard Quality Information model for competitive advantage in business.

Analysis and design of Balanced Scorecard Quality Information for competitive advantage in the automotive industry of Thailand is a method of quality information technology and a new discipline of knowledge development by integrating information technology science and quality methodology with the main sciences to solve the problems of working or enhance the performance of various organizations.

So that information can be valuable for use in various works, information must be of quality acceptable to general users. The method: QIT = Quality Information Technology or quality information technology is based on the integration of at least 3 related disciplines of quality, causing innovations of higher quality comprising: 1. the main discipline, i.e. business administration; 2. quality methodology science, namely Balanced Scorecard quality methodology based on 4 perspectives on business success as follows: [2] financial perspective; customer perspective; internal process perspective as well as learning and growth perspective. 3. Information technology science, i.e. information technology tools such as analysis and design of Stepwise Refinement system of information will encourage the effective integration of all 3 disciplines. This allows us to obtain Balanced Scorecard Quality Information for competitive advantage in business, which can be further developed into software in the future as shown in Fig. 1.

Fig. 1. Concept of quality information development according to quality information theory

The above picture shows that the Balanced Scorecard quality methodology was used to increase IT quality, i.e. analysis and design of information to support business competition using Balanced Scorecard Quality Information (BSCI) designed, which can be further developed into software (CAQI). The clear concept can be viewed and shown in Fig. 2 [3].

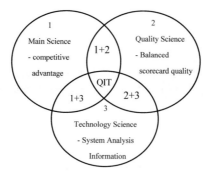

Fig. 2. Concept of quality information development Preliminary

2 Preliminary

2.1 Balanced Scorecard

Balanced Scorecard Concept It is a concept that creates value for the business. It helps in measuring corporate performance from corporate executives. Companies use this tool to build a business future that is used to link short-term controls. And long-term organizational alignment with the corporate vision, which consists of various perspectives of the organization. All four aspects are related to time dimension, Past, present and future In addition to being a measure of performance. It is also a tool to drive the performance of the organization in various areas as well.

2.2 Quality Information Technology

Quality Information Concepts It brings the concept of quality science. To integrate with the concept of information technology. And then it creates a new tool that is characterized by quality information tools. And can be used to develop quality in various organizations.

IT =   information technology
QM =   quality methodology
QIT =   quality information technology
CAQI =   computer – aided quality improvement

2.3   Business Intelligence Dashboard

A business intelligence dashboard. It is a data visualization tool that displays the current status of your organization's core performance metrics. The dashboard displays performance indication for various aspects of the organization. Business intelligence dashboards consist of various interfaces that represent real-time data from multiple sources.

2.4   Review Literature and Related Research

Abbas Asosheh, Soroosh Nalchigar and Mona Jamporazmey [4] Information technology (IT) is a tool crucial for enterprises to achieve a competitive advantage and organizational innovation. A critical aspect of IT management is the decision whereby the best set of IT projects is selected from many competing proposals. The optimal selection process is a significant strategic resource allocation decision that can engage an organization in substantial long-term commitments. However, making such decisions is difficult because there are lots of quantitative and qualitative factors to be considered in evaluation process. This paper has two main contributions. Firstly, it combines two well-established managerial methodologies, balanced scorecard (BSC) and data envelopment analysis (DEA), and proposes a new approach for IT project selection. This approach uses BSC as a comprehensive framework for defining IT projects evaluation criteria and uses DEA as a nonparametric technique for ranking IT projects. Secondly, this paper introduces a new integrated DEA model which identifies most efficient IT project by considering cardinal and ordinal data. Applicability of proposed approach is illustrated by using real world data of Iran Ministry of Science, Research and Technology.

Rajesh, Pugazhendhi, Ganesh, Ducq and Lenny Koh [5] To provide valuable support for successful decision-making, managers needs a balanced set of financial and non-financial measures that represent different requirements, strategic goals, strategies, resources, and capabilities and the causal relationships between these domains. The balanced scorecard (BSC) is such a measurement system. Although much discussion has taken place in industries and academia circles for the development of BSC for third party logistics (3PL) service provider, little research exists which studies and develops BSC strategies for 3PL service providers. This study proposed a set of strategies for BSC of 3PL service providers. We devised a strategies framework for all the four BSC perspectives of the various functions of 3PL service providers and the weightages for the different strategies are evaluated using Delphi analysis. The implementation of the proposed framework in a 3PL company is also discussed.

Kadarova, Durkacova and Kalafusova [6] In recent years academic scholars have given increasing attention to the importance of strategic measurement systems including both non-financial and financial measures and have focused attention on the method called Balanced Scorecard. Through the Balanced Scorecard, an organization monitors both its current performance (finance, customer satisfaction, and business process results) and its efforts to improve processes, motivate and educate employees, and enhance information systems – its ability to learn and improve.

Amado, Santos, Marques [7] This article presents the development of a conceptual framework which aims to assess Decision Making Units (DMUs) from multiple perspectives. The proposed conceptual framework combines the Balanced Scorecard (BSC) method with the non-parametric technique known as Data Envelopment Analysis (DEA) by using various interconnected models which try to encapsulate four perspectives of performance (financial, customers, internal processes, learning and growth). The practical relevance of the conceptual model has been tested by using it to assess the performance of DMUs in a multinational company which operates in two business areas. Various models were developed with the collaboration of the directors of the company in order to conceive an appropriate and consensual framework, which may provide useful information for the company. The application of the conceptual framework provides structured information regarding the performance of each DMU (from multiple perspectives) and ways to improve it. By integrating the BSC and the DEA approaches this research helps to identify where there is room for improving organisational performance and points out opportunities for reciprocal learning between DMUs. In doing so, this article provides a set of recommendations relating to the successful application of DEA and its integration with the BSC, in order to promote a continuous learning process and to bring about improvements in performance.

Jalaliyoon, Bakar, Taherdoost [8] The value of operation assessment has been specified for organizations and it plays a vital role in most of the organizations. Complexity and frailty of decision making in business makes strategic management imperative. On the other hand, key operational factors are fiscal and non-fiscal measurement criteria used to identify goal's quality and reflect an organization strategic action. These factors are used to assess the present condition of the company and define suitable solutions to business methods. This study proposes an appropriate methodology in order to design and implement balanced score card (BSC) for operational appraisal of industrial groups. In order to achieve this goal thirteen required steps is proposed and discussed.

3   Proposed Method

3.1   Balanced Scorecard Quality Information Dashboards Model
for Competitive Business Advantage

Analysis and design of Balanced Scorecard Quality Information for competitive advantage in business consisting of:

1. Study of the necessary use of the users by classifying information according to the work, [9] which can be divided into 3 levels, comprising senior management, intermediate management and operational management (Fig. 3)
2. Study of job details of each position
3. Creation of a Balanced Scorecard Quality Information analysis table based on the work of the management at different levels in 4 perspectives according to Balanced Scorecard quality methodology to generate information supporting the work of each level (Table 1).
4. Design of Balanced Scorecard Quality Information

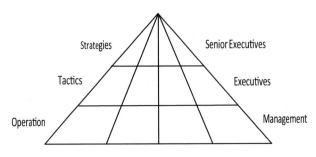

Sales Production Finance & Account Human Resources

Fig. 3. Types of information systems and management levels in organizations

Table 1. Balanced Scorecard Quality Information analysis table based on the work of the management

Management level	Finance				Customer			
Senior executives	FH1	FH2	FH3	FH4	CH1	CH2	CH3	CH4
Executives	FM1	FM2	FM3	FM4	CM1	CM2	CM3	CM4
Management	FO1	FO2	FO3	FO4	CO1	CO2	CO3	CO4
Management level	Internal process				Learning and growth			
Senior executives	IH1	IH2	IH3	IH4	LH1	LH2	LH3	LH4
Executives	IM1	IM2	IM3	IM4	LM1	LM2	LM3	LM4
Management	IO1	IO2	IO3	IO4	LO1	LO2	LO3	LO4

5. Use of stepwise refinement method, analysis of relationships and completeness of information of each issue, resulting in 66 items of information.
6. Analysis of the validity of information content, resulting in the reliability equaling 0.94.

$$\text{Reliability} = \frac{2M}{N_1 + N_2}$$
$$\text{Reliability} = \frac{2(124)}{132 + 132}$$
$$= 0.94$$

7. Assessment of acceptance of Balanced Scorecard Quality Information for competitive advantage in business.

4 Experimental Results

The research results can be explained as follows.

1. Getting the Balanced Scorecard Quality Information model for competitive advantage in business (66 items), being divided into 8 items for senior management, 20 items for intermediate management and 38 items for operational management (Tables 2, 3, 4 and 5).

Table 2. Table showing Balanced Scorecard Information from financial perspective classified by management level

Organizational level	Financial aspect					
	Determination of policies and strategies	Targeting	Data analysis	Reporting and checking	Administration and controlling	Performance evaluation
Senior executives	1				1	
Executives			2	3	1	
Management				4	4	

Table 3. Table showing Balanced Scorecard Information from customer perspective classified by management level

Organizational level	Customer aspect					
	Determination of policies and strategies	Targeting	Data analysis	Reporting and checking	Administration and controlling	Performance evaluation
Senior executives	1				1	
Executives			2	1	1	
Management			2	5	1	

Table 4. Table showing Balanced Scorecard Information from internal process perspective classified by management level

Organizational level	Internal process					
	Determination of policies and strategies	Targeting	Data analysis	Reporting and checking	Administration and controlling	Performance evaluation
Senior executives	1				1	
Executives			1	1	2	1
Management			2	3	4	2

Table 5. Balanced Scorecard Information Table from learning and growth perspective classified by management level

Organizational level	Learning and growth					
	Determination of policies and strategies	Targeting	Data analysis	Reporting and checking	Administration and controlling	Performance evaluation
Senior executives	1				1	
Executives			3	1	1	
Management			3	3	5	

2. The content of Balanced scorecard Quality information dashboards model for competitive advantage in business is valid in accordance with Balanced Scorecard methodology and business competition with the reliability of expert analysis at 0.94 level.
3. The relevant stakeholders accepted Balanced Scorecard Quality Information for competitive advantage in business designed. The number of accepting persons was higher than the denying persons in a statistically significant manner at 0.05 level.
4. Balanced Scorecard Quality Information is displayed in the form of Dashboard to help in decision making (Figs. 4, 5 and 6).

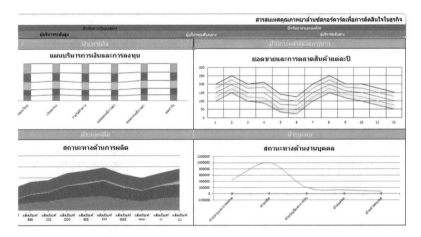

Fig. 4. Balanced Scorecard Quality Information dashboard for senior executives level

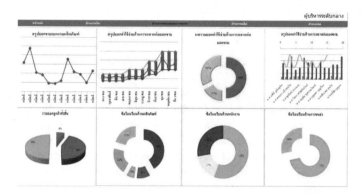

Fig. 5. Balanced Scorecard Quality Information dashboard for executives level

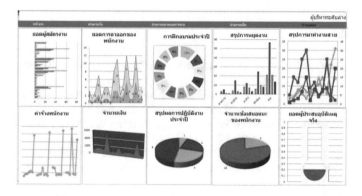

Fig. 6. Balanced Scorecard Quality Information dashboard for management level

5   Conclusion

This report proposes the Balanced Scorecard Quality Information display model was the process for an organization to develop quality information model using the integration of the QIT analysis and design method with the stepwise refinement algorithm. This integration methodology could be useful to analyze and design quality information in other organizations. It was found that overall the senior management accepted Balanced Scorecard Quality Information display model for competitive advantage in business, totaling 8 items, the intermediate management, totaling 20 items and the operational level, totaling 38 items. The values obtained were calculated, resulting in the following averages of 3.88, 4.0 and 4.02, respectively, meaning that the information display model was accepted at a high level. When considering the individual items, it was found that the top three accepted information of each level is as follows: senior management, i.e. sales and marketing plan, financial and investment plan, personnel management plan; intermediate management, i.e. summary of the total customers, summary of the total marketing expenses per sales, production capacity, cost of production, balance sheet, cash position of business, staff satisfaction, payroll summary; operational management, i.e. summaries of order cancellation, monthly marketing expenses, sales classified by staff, the request, issue of inventory, production schedule, the total monthly consumption of raw materials, the total investment, the total loans of business, the number of accounts receivable per customer, the total resignation of staff, summary of the total leave, overtime pay.

The content of Balanced Scorecard Quality Information display model for business competitive advantage is valid according to Balanced Scorecard methodology and business competition with the reliability of expert analysis at 0.91 level.

The management at each level in accordance with the concept of Lon accepted Balanced Scorecard Quality Information for business competitive advantage in terms of competitive advantage, including: (1) cost advantage (2) difference advantage (3) rapid response advantage and accepted various aspects, namely (1) simplicity (2) understanding (3) usefulness (4) accuracy (5) image (6) the quality of being economical.

References

1. Sungrung, P.: Production management. Suan Dusit Rajabhat University, Bangkok, Thailand (2003)
2. Kaplan, R.S., Norton, D.P.: The balanced scorecard: translating strategy into action. Publications the President and Fellows of Harvard College (1996)
3. Chookittikul, J.: Quality information technology. Computer and Advanced Technology, 8th edn., vol. 8, pp. 8–9. Phetchaburi Thailand (2005)
4. Asosheh, A., Nalchigar, S., Jamporazmey, M.: Information technology projectevaluation: An integrated data envelopment analysis and balanced scorecard approach. Expert Syst. Appl. **37** 5931–5938 (2010).
5. Rajesh, R., Pugazhendhi, S., Ganesh, K., Ducq, Y., Lenny Koh, S.C.: Generic balanced scorecard framework for third party logistics service provider. Int. J. Prod. Econ. **140**, 269–282 (2012)
6. Kadarova, J., Durkacova, M., Kalafusova, L.: Balanced scorecard as an issue taught in the field of industrial engineering. Procedia Soc. Behav. Sci. **143**, 174–179 (2014)
7. Amado, C.A.F., Santos, S.P., Marques, P.M.: Integrating the data envelopment analysis and the balanced scorecard approaches for enhanced performance assessment. Omega **40**, 390–403 (2012)
8. Jalaliyoon, N., Bakar, N.A., Taherdoost, H.: Propose a methodology to implement balanced scorecard for operational appraisal of industrial groups. In: The 7th International Conference Interdisciplinary in Engineering, vol. 12, pp. 659–666 (2014)
9. Laudon, K.C., Laudon, J.P.: Management Information Systems: Managing the Digital Firm, 9th edn. Person Prentice Hall, Upper Saddle River (2006)

Criminal Background Check Program with Fingerprint

Narumol Chumuang[1(✉)], Mahasak Ketcham[2],
and Amnat Sawatnatee[3]

[1] Department of Digital Media Technology, Faculty of Industrial Technology
Muban, Chombueng Rajabhat University, Chom Bung, Thailand
lecho20@hotmail.com
[2] Department of Information Technology Management, Faculty of Information
Technology, King Mongkut's University of Technology North Bangkok,
Bangkok, Thailand
mahasak_k@it.kmutnb.ac.th
[3] Department of Multimedia Technology, Faculty of Science, Chandrakasem
Rajabhat University, Chom Bung, Thailand

Abstract. Fingerprint recognition is one of the most widely used methods of biometrics. This method relies on the surface topography of a finger and, thus, is potentially vulnerable for spoofing by artificial dummies with embedded fingerprints. In this study, we applied the fingerprint recognition technique to check criminal background of person. Our system working with two parallel process that are fingerprint recognition and matching with personal profile in database. We also demonstrated that an evaluated the experimental with two types that are match and not match to any profile in database which can used in automatic recognition systems. The efficiency of our system is excellent.

Keywords: Fingerprint · Check · Background · Criminal

1 Introduction

Personal identification there are many ways to verify and prove. Whether prove by ID card or official document are issue by government, such as a fingerprint, signature, etc. The government is more important to identify who person and will be complicated and costly. The tools or equipment belong to government is not up to date and low technology.

Fingerprints are another way to identify [1] a person and are not complicated, not need to by document, Even though the fingerprint reader tools is not expensive but cannot be read or process by itself and not to have other device to process shut as computer. The Fingerprint Imaging Program for personality identification [2] is a concept to minimize the work process and easy to use and user friendly. The Personality identification process can be made easier and more convenient by using Minutiae matching.

© Springer Nature Switzerland AG 2019
T. Theeramunkong et al. (Eds.): iSAI-NLP 2017, AISC 807, pp. 169–186, 2019.
https://doi.org/10.1007/978-3-319-94703-7_16

2   Related Literature

2.1   Fingerprints

Fingerprints Feature and specifications. The Fingerprints are often used to identify or prove to be the same person. The Fingerprints is the most reliable way and unique but there is little change depend on age [3]. Fingerprints consist of two types of stripes. The first called "Ridge" caused by the embossed surface from the outer skin. And another type called "Groove or Valley" is a depth that is lower than the level of the convex. So you will see that the skin on the finger will have a higher line and another line deep. The convex lines and grooves are alternate [3] (Fig. 1).

Ridge

Groove or
Valley

Fig. 1. Groove and the bulge of the finger.

Fingerprints can be classified into four patterns: Arch, Ulnar Loop, Whorl by have detail as:

1. Arch patterns: This is a special pattern and unique. It only 5% of person for Arch pattern and can be define; (1) simple arch, (2) tented arch (Fig. 2).
2. Loop patterns: This pattern is most common of fingerprints and can be founded 60–65% and can be define as (1) ulnar loop, (2) radial loop (Fig. 3).
3. Whorl pattern is a fingerprint with a circular line around the middle, similar to aspiral. The person with this fingerprint pattern is about 30–35% and can be define concentric whorl and whorl spiral.
4. Mixed Fingerprints is a special fingerprint and cannot be matched to any of the three types mentioned above, but may consist of two fingerprints mixed and more than two shoals or more [4] (Fig. 4).

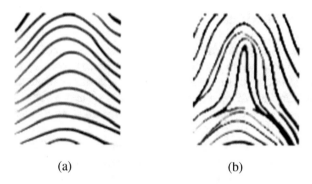

(a)   (b)

Fig. 2. Arch patterns (a) simple arch, (b) tented arch.

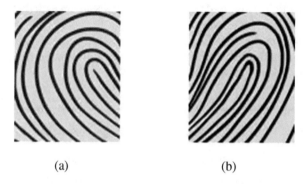

(a)   (b)

Fig. 3. Loop patterns (a) ulnar loop and (b) radial loop.

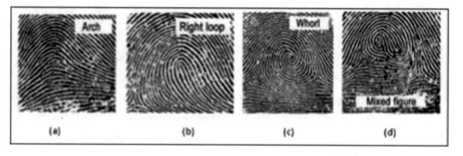

Fig. 4. Fingerprint patterns (a) Arch (b) Right loop (c) Whorl (d) Mixed figure.

2.2   Matching the Fingerprint

Fingerprint Matching Methods are fingerprint or finger scan techniques that compare the registered fingerprint templates. With fingerprints being scanned as fingerprint matching, the techniques commonly used in finger scanners present two types of matching: miniature matching and pattern matching [5].

2.2.1   Miniature Matching

This method uses the principle that each one of us fingerprints. The line form is often referred to as the junior point. Bifurcation is the point where the line separates and the endings are the places where the line ends. In the finger recording process for registration. Miniature point are generated in the form of positions relate to each other both of position and direction. This will be recorded as a reference value and converted to digital data stored as a template. This will be recorded as a reference value and converted to digital data stored as a template.

This data is stored as the value of the underlying fingerprint. To be used to monitor the person when there is comparison of usage Miniature matching occurs when a finger scan of a fingerprint user is performed. Of users who have been added. It will undergo a mini checkout process as well as a master fingerprint image.

The reference points that have been separated as the junctions are converted to digital data and will be compared the level of similarity to the underlying image. It was recorded at the time of registration.

By trying to compare points. As much as possible in value definitely defined. This value will determine the comparative results obtained. It must be close to how well it will match as a junior as shown in Fig. 5 [6].

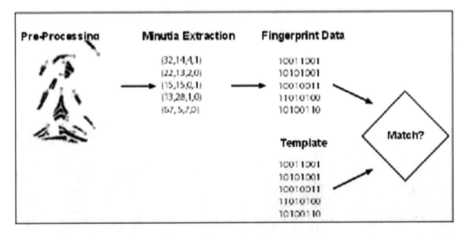

Fig. 5. Minutiae matching process.

2.2.2   Pattern Matching

One unique feature of algorithm the pattern is the effect of all fingerprints. Counted not only for specific fingerprint effects. Like a mini match. Include sub areas, this includes the thickness, the curvature, and density of the lines [7, 8]. The reason of these is the complexity of the algorithm. The database model is independent of the size of the fingerprint sensor and does not depend on the capacity.

Minutiae in Fingerprints

To the extent of the way the Minuses. It will be damaged by the difficulty of finger recognition with variable fingerprints. But for matching algebraic the format is not.

So the graphics that come from the capture device to distinguish it from those stored in the database. The operation of the program to determine fingerprints and centering, which may not be the center of gravity from middle finger.

After that, the image is cut at a fixed distance around the center of the image. It is a rectangle, after which the cut area is compressed and stored for later pairing. The proof process starts with a fingerprint image of the user. After that, the prototype image recorded at registration will be compared to the scanned fingerprint image. To determine the difference with the underlying image, a Verification Threshold value is used, which describes the minimum permissible deviation that is used to determine the fingerprint. Match the same as the one stored as shown in Fig. 6 shown a fingerprint pattern matching.

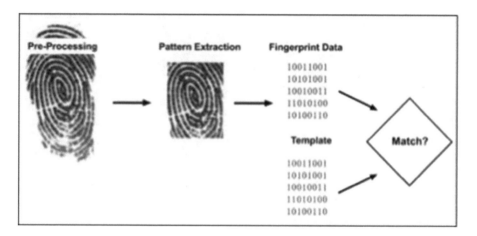

Fig. 6. Illustration a fingerprint pattern matching

3   Proposed Method

In this section, the detail for checking criminal background with fingerprint are described. Start with system overview as Fig. 7 then followed by the sub-process detail. From Fig. 7, the program has two parts. First is the fingerprint storage to database process and the other one is the fingerprint verification process.

By the way, both of part are a similar process. The fingerprint storage process will be received fingerprint image from the criminal record and to improve fingerprints and validate key point and convert the data to storing.

After that next process is fingerprint validation, this process will be printed the fingerprint to smooth surface objet, white colure non texture then capture by cell phone's camera and checking the fingerprint identity from database.

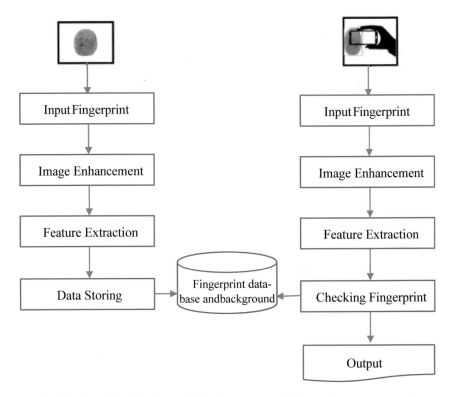

Fig. 7. The criminal background check program with fingerprint system overvie

3.1   Fingerprint Enhancement

The fingerprint quality improving is a step to make the image sharpness and better and makes the result higher accuracy. Because some part of the image has lost or interference from outside environment, while shooting or blocked by oil of the skin or fingerprint, was over-ink printed. So, the image will not be perfected then the fingerprint quality Improving is the one to improve image as a basis, There are 5 steps as:

3.1.1   The Image Adjusting to Gray Tone Procedure
When we import fingerprint image from cell phone camera the image has blue tone and to make easier improving process will be change to gray scale. The gray image intensity is depend on the number of bits used. For example, an 8-bit grayscale image at a total 256 color level, typically in the 0–1 or 0–255 range.

3.1.2   Image Intensification Procedure
After adjust the image to Grayscale the next step will be sharpness of the image. Because the image might loss some details in Grayscale conversion process. So, the Picture intensification process will improve image details, lines and edges.

3.1.3   The Monochrome Image Adjusting Procedure

After adjusting the image to grayscale and contrast adjustment which state by digital has only 2 states

(1) 0: Means that the pixel is black
(2) 1: Means that the pixel is white It will easy to identify to the next feature.

3.1.4   Data Storage Procedures

It will be done after the monochrome image adjusting procedure successfully. It will save the image to database and takes the fingerprint image for authentication (Figs. 8 and 9).

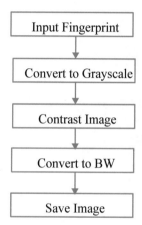

Fig. 8. Process of image enhancement diagram

3.2   Fingerprint Characterization Process

Characterization is a process of finding the position of a muzzle, which is divided into two types. (1) End Point (Ridge Ending) and (2) Split Point (Ridge Bifurcation) which procedure is developing based on the workflow for finding the number of pixels in each pixel. By start checking the gray of each pixel where value is around 255 and if found pixel value is 255 surrounding area as equal to 1 or 3 that pixel is the end point or points separated after that will keep that value, position by row and column at that pixel stored to the buffer. MINUTIAE will transfer to Minutiae regulation process by position correcting. The principle is to find distance between the two nodes, where the distance is stored in a row and column. Between the two points is less than 4 that mean two MINUTIAE are close together and one of miniature will be removed [1].

3.3   Comparison of Identity Points

Fingerprint authentication is a microscopic method of checking fingerprints. The number of databases that are included in the core database is minutia alignment and

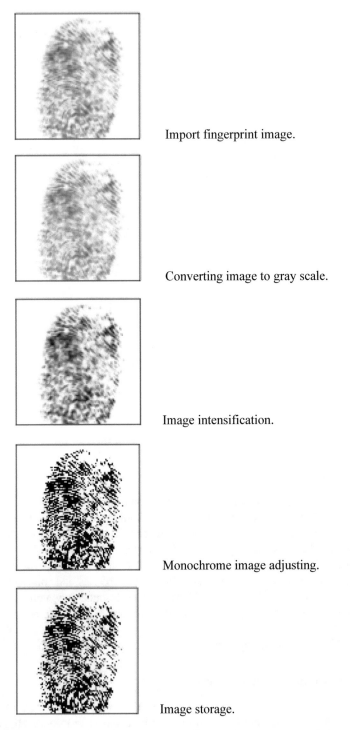

Import fingerprint image.

Converting image to gray scale.

Image intensification.

Monochrome image adjusting.

Image storage.

Fig. 9. The process of improving the quality of fingerprint images diagram.

minutia matching. Minutia is the finding of fingerprints that needs to be monitored and consistent. With the structure around the muzzle in the database of 5 points after the successful adjustment. This is a pairwise process where the principle is to count the number of pairs of fingerprints matching the fingerprint of the fingerprint in the database. Based on the distance from each point. The result will be in percentage format. If the match is less than 70%, it is not the same fingerprint. But if more than or equal Considered to be the same fingerprint [1].

The distance will check between five 5-node of Minutia and compared to the input reference, defining ε as the following tolerances as: imported image with the same images in the database, the average is 1 or closed to 1.

The image value x of axis - (import value x axis - reference value x axis) <= 0 or, the image value y of axis - (Import value y axis - reference value y axis) <= 0 from above eq, all of points will be measured 5 times then will average the measuring value each time (Fig. 10).

Fig. 10. The dots color comparison as Minutiae method diagram.

The image shows the color dots that has calculated values which the computation takes the intersection point and the computed point. Minutiae comparisons.

3.4   Working Process Output Program

The output of the program will compare fingerprint input data and database. If the image are matched to data, will be displayed along with the name and crime information and if the data does not match. The program will not display (Figs. 11 and 12).

4   Experimental and Results

This work is comparing the fingerprints that was taken from mobile phone camera. The object was printed on white paper and no texture. The process has three steps (1) improve quality of images, (2) identify and (3) fingerprint authentication.

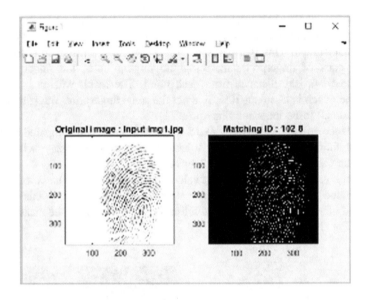

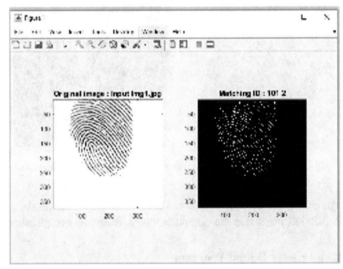

Fig. 11. The output where fingerprint image is matching.

We have taken photograph of 20 fingerprints and divided into fingerprints in 10 databases and fingerprints that were not in the database. It is brought into the process of validation and comparison to test data consistency. The results of classification is separated into three categories as follows.

1. Found fingerprints and found criminal records.
2. Found fingerprints, but no criminal records.
3. Fingerprints not found.

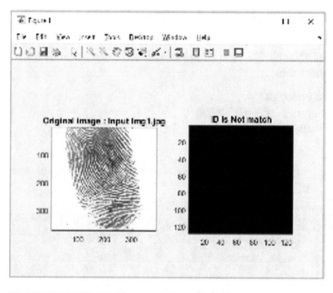

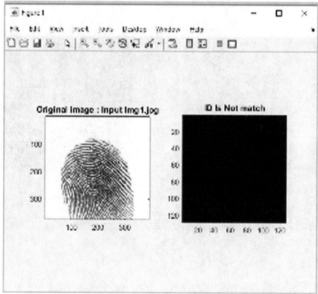

Fig. 12. The output where fingerprint image is not match

The results shown as 20 images of fingerprints was processed accurately. In case matches or unmatched was showed 100% accuracy and timing was processing 1–2 min per process. In the future, the researcher has proposed that this program can be developed and access real-time to the database and developed to mobile application.

4.1 Case Fingerprints and Criminal History Findings

Result: Case 1

The results showed the input image is matched to database as ID 102_4.tif, compared to database tables.

Fingerprint code: 102_4
Identification no. 6683463048306
Name - Family name Mrs Uthid Pakdeesuk
Lawsuits 2

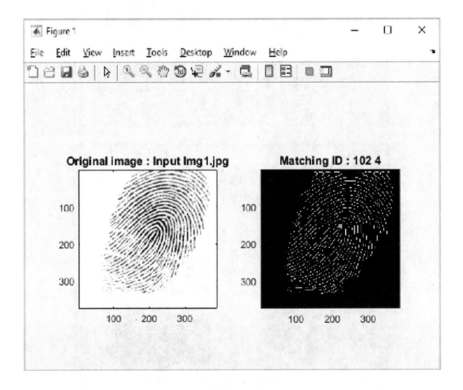

Result: Case 2

Command Code: Computing similarity between Input Img1.jpg and 102_8 from FVC2002: 1

The results showed the Input Img is matched to database as ID 102_5.tif, compared to database tables.

Fingerprint code: 102_5
Identification no.: 2291488490915
Name - Family name Mr Kreangsak Suksrikeaw
Lawsuits 3

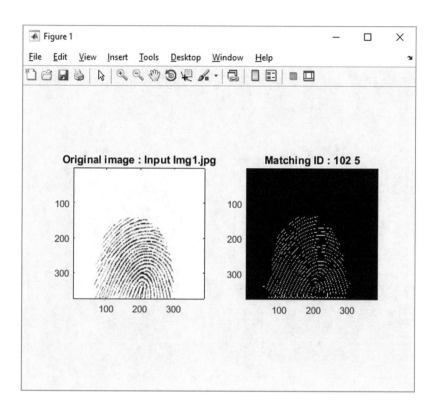

Result: Case 3

Command Code: Computing similarity between Input Img1.jpg and 102_6 from FVC2002: 1

The results showed the Input img is matched to database as ID 102_6.tif, compared to database tables.

Fingerprint code: 102_6
Identification no.: 8310401127322
Name-Family name Mrs Ratchaneekorn Huedkhuntod
Lawsuits 3

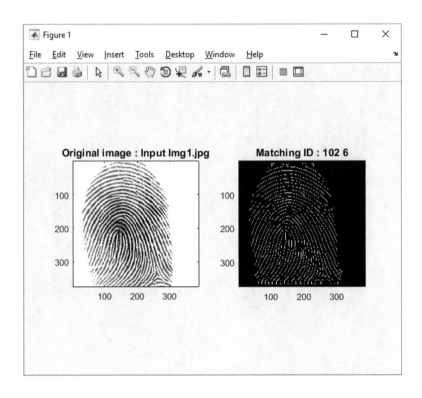

Result: Case 4

Command Code: Computing similarity between Input Img1.jpg and 102_8 from FVC2002: 1
Output 4

The results showed the Input Img is matched to database as ID 102_7.tif, compared to database tables.

Fingerprint code: 102_7
Identification no.: 5047941774543
Name-Family name Miss Yamaporn Kohkerd
Lawsuits 1

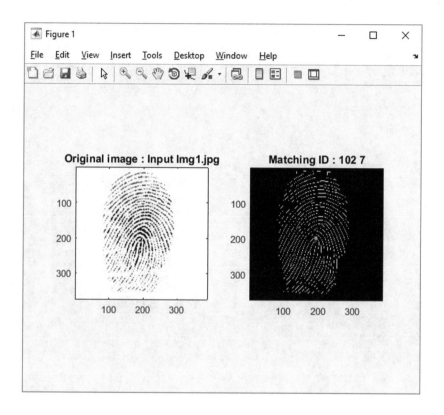

Result: Case 5

Command Code: Computing similarity between Input Img1.jpg and 102_8 from FVC2002: 1
Output 5

The results showed the Input Img is matched to database as ID 102_8.tif, compared to database tables.

Fingerprint code: 102_8
Identification no.: 3814129835766
Name-Family name Mr.noppadol suamuangpan
Lawsuits 2

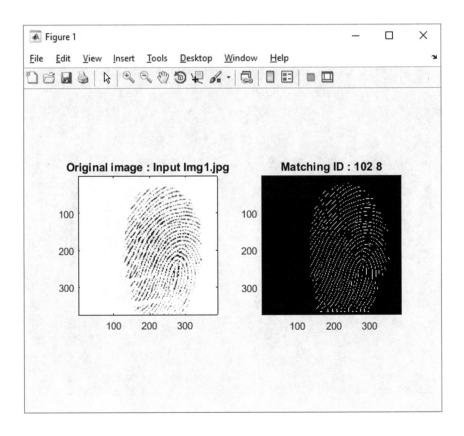

4.2   In Case Fingerprint Image Was not Found Criminal Record

Result: Case 6

Command Code: Computing similarity between Input Img1.jpg and 102_8 from FVC2002: 1
Output 6

The results showed the Input Img is matched to database as ID 101_1.tif, compared to database tables.

Fingerprint code: 101_1
Identification no.: 2190116378731
Name-Family name Miss Sayfon Jananchupol, Not found criminal record.

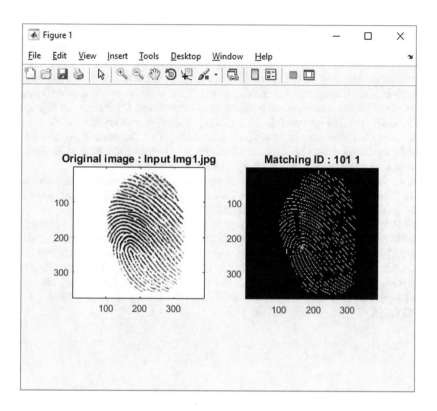

5   Conclusion

The performance of checking criminal background profile with fingerprint systems is based on image processing. In this paper present a highlights of the event to useful methodology for fingerprint crime recorder can be used to assist in criminal history investigations. For helping to reduce the cost and time spent investigating crime records. Moreover, the fingerprint verification program can be used to develop a fingerprint verification program cell phone. Which can be used in the real world and increased security for life and property.

References

1. Fujun, F., Xinshe, L., Litao, W.: Design and implementation of identity authentication system based on fingerprint recognition and cryptography. In: 2016 2nd IEEE International Conference on Computer and Communications (ICCC), Chengdu, China, pp. 254–257 (2016). https://doi.org/10.1109/compcomm.2016.7924704
2. Soewito, B., Gaol, F.L., Simanjuntak, E., Gunawan, F.E.: Smart mobile attendance system using voice recognition and fingerprint on smartphone. In: 2016 International Seminar on

Intelligent Technology and Its Applications (ISITIA), Lombok, pp. 175–180 (2016). https://doi.org/10.1109/isitia.2016.7828654

3. Ali, M.M.H., Mahale, V.H., Yannawar, P., Gaikwad, A.T.: Overview of fingerprint recognition system. In: 2016 International Conference on Electrical, Electronics, and Optimization Techniques (ICEEOT), Chennai, pp. 1334–1338 (2016)

4. Hnoohom, N., Chumuang, N., Ketcham, M.: Thai handwritten verification system on documents for the investigation. In: 2015 11th International Conference on Signal-Image Technology and Internet-Based Systems (SITIS), Bangkok, pp. 617–622 (2015)

5. Chumuang, N., Ketcham, M.: Intelligent handwriting Thai Signature Recognition System based on artificial neuron network. In: TENCON 2014 – 2014 IEEE Region 10 Conference, Bangkok, pp. 1–6 (2014). https://doi.org/10.1109/tencon.2014.7022415

6. Lin, C., Kumar, A.: Improving cross sensor interoperability for fingerprint identification. In: 2016 23rd International Conference on Pattern Recognition (ICPR), Cancun, pp. 943–948 (2016). https://doi.org/10.1109/icpr.2016.7899757

7. Patil, A.P., Bhalke, D.G.: Fusion of fingerprint, palmprint and iris for person identification. In: 2016 International Conference on Automatic Control and Dynamic Optimization Techniques (ICACDOT), Pune, pp. 960–963 (2016). https://doi.org/10.1109/icacdot.2016.7877730

8. Yani, W., Zhendong, W., Jianwu, Z., Hongli, C.: A robust damaged fingerprint identification algorithm based on deep learning. In: 2016 IEEE Advanced Information Management, Communicates, Electronic and Automation Control Conference (IMCEC), Xi'an, pp. 1048–1052 (2016)

Jobs Analysis for Business Intelligence Skills Requirements in the ASEAN Region: A Text Mining Study

Waranya Poonnawat[1(✉)], Eakasit Pacharawongsakda[2], and Nuttaporn Henchareonlert[1]

[1] School of Science and Technology, Sukhothai Thammathirat Open University, Nonthaburi, Thailand
{waranya.poo,nuttaporn.hen}@stou.ac.th
[2] Big Data Engineering Program, College of Innovative Technology and Engineering, Dhurakij Pundit University, Bangkok, Thailand
eakasit.pac@dpu.ac.th

Abstract. This paper applies text mining and data mining techniques to analyse online job advertisements related to Business Intelligence (BI) in web portals around the ASEAN region. Several techniques are applied: data preprocessing, tokenization, association rules (AR) with FP-Growth and visualisation. The research goal is the profiles of BI skills required for industry and the market. The results would be useful for individual to prepare for their BI competence, for faculties to prepare for BI-related courses and curriculum, and for organizations to prepare for Human Resource (HR) and job advertisement purposes.

Keywords: Text mining · Business intelligence · Skills · Jobs

1 Introduction and Statement of Problem

Over the past two decades, academics and practitioners have been paying attention to Business Intelligence (BI) [1]. Business demand that can be seen from the CIO survey report showed that BI is one of the top five from the priority technologies and has been for several years, and it is applied to all business segments [1, 2]. Universities have also responded to the industry both in educational and research demands [1]. BI-related courses are integrated in several curriculums and levels worldwide [3]. BI is considered as an interdisciplinary area, which covers such areas as database systems, data warehouse, data mining, text mining, network analysis, etc. However, BI skills do not only cover technological skills and analytical skills, but also the demand for business knowledge and communication skills, which is the ability to understand business, interpret the analytic results and explain to business people [1]. Moreover, knowledge and skills in BI is a foundation for Information Systems (IS) workers, and is still important for developing the skills in advanced technology [3].

The gap between the BI skills needed by industries and BI skills provided for graduates from universities needs to be addressed, and it is an essential concern for all faculties. Additionally, the requirements of BI skills of industries have changed periodically

© Springer Nature Switzerland AG 2019
T. Theeramunkong et al. (Eds.): iSAI-NLP 2017, AISC 807, pp. 187–195, 2019.
https://doi.org/10.1007/978-3-319-94703-7_17

because of the revolution of technology, and more challenges have increased such as business analytics have changed from descriptive analytics to more predictive analytics and prescriptive analytics. The data structure has changed from structured data to more semi-structured data and unstructured data. Moreover, BI technology is dynamic and highly competitive, therefore, BI workers need to be aware of the trend and the skills needed by the market and industry.

This research would like to extract, identify, categorise and highlight the skills needed for BI-related jobs from online jobs portals in ASEAN countries (i.e., Brunei Darussalam, Cambodia, Indonesia, Lao PDR, Malaysia, Myanmar, the Philippines, Singapore, Thailand and Vietnam). This would be useful for faculties and universities to prepare for improving or developing BI-related courses and curriculum, and for students to prepare for the BI skills needed by industry.

The objectives of this research are as follows: (1) to review and extract online job posts related to BI from popular job portals in the ASEAN region; (2) to analyse and categorise the skills needed for BI-related jobs, and (3) to visualise the taxonomy and profiles of BI-skills needs of the industry.

2   Literature Review

2.1   Business Intelligence

Hans Peter Luhn [4] first introduced the term of "Business Intelligence" in October 1958 in an IBM journal. BI was described as an automatic system that can accept information, disseminate the information properly and furnish information on demand. Grimes [5] mentioned that this vision was before the emergence of computerised business operations and decision support technology, however, todays direction of BI systems focuses on the automated knowledge analytics that was conceptualised by Luhn.

Power, who explored the developments in DSS between 1952 and 2002, stated that generally BI systems are data-driven DSS [6]. A data-driven DSS with emphasis on access and manipulation of the time-series internal and external data can be broadened by means of data warehouse technology for data manipulation and OLAP technology for data analysis [6].

The term BI has become popular since 1989 because Howard Dresner from the Gartner Group promoted it, and coined it as an umbrella term for describing *"concepts and methods to improve business decision-making by using fact-based support systems"* [7, p. 1]. BI has also been used to describe analytical and decision support applications, leading to a broader definition of BI. Wixom and Watson [8, p. 14] defined BI as *"a broad category of technologies, applications, and processes for gathering, storing, accessing, and analysing data to help its users make better decisions"*. Chiang, Goes and Stohr [9, p. 12:2] defined BI by combining Business Intelligence with Business Analysis (BI&A) and stated that *"BI&A is used as a unified term to describe information-intensive concepts and methods to improve business decision-making...and includes the underlying architectures, analytical tools, database management systems, data/text/web mining techniques, business applications, and methodologies"*. The authors also state that BI&A is an interdisciplinary area that requires three areas of knowledge and skills for effective BI&A, namely: analytical skills (e.g., data mining, sentimental analysis, statistical analysis, text mining); Information Technology (IT) knowledge and skills (e.g., relational database, data mart and data warehousing, ETL operations, OLAP, semi-structured and unstructured data management, cloud computing); and business knowledge and communication skills (e.g., understanding the business issues, framing the analytical solutions, explaining things in simple terms, broadcasting the lessons learned). Moreover, BI Skills are highly important for Information Systems (IS) students in modern dynamic business environments.

2.2   Text Mining

Text Mining is known as text data mining or knowledge discovery from textual databases and refers to the process to extract any interesting patterns or knowledge from text documents [10]. A general framework (Fig. 1) of text mining consists of two phases: text refining and knowledge distillation. Text refining transforms free-form text documents into an Intermediate Form (IF) which can be document-based (each entity represents a document) or concept-based (each entity represents an object of interest in a specific domain). The mining operations for document-based form are those such as clustering and categorisation, and for concept-based those such as predictive modelling and associative discovery. Additionally, a document-based form can be transformed into a conceptbased form.

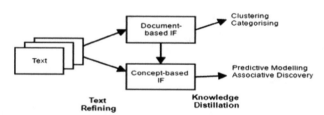

Fig. 1. A text mining framework (Source: Tan 1999, p. 2)

Cai and Sun [11] also state that text mining is a kind of data mining to discover hidden knowledge and patterns from text data in heterogeneous sources.

The process of text mining (Fig. 2) as follows: (1) data selection selecting proper data to be processed and analysed; (2) data preprocessing filtering any noise data and performing any preliminary processes; (3) data transformation converting to the format suitable to be processed by mining algorithms; (4) text mining applying text mining algorithms to find patterns and knowledge, and (5) interpretation or evaluation evaluating the patterns and knowledge which are interesting and have the most accurate model.

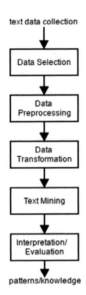

Fig. 2. Knowledge discovery using data mining (Source: Cai and Sun 2016, p. 1)

2.3 Related Works

Litecky et al. [12] analyses the web content using text mining application for skills related to Information Technology (IT) jobs. The authors developed a software to search job advertisement daily from three websites (i.e., Monster.com, HotJobs.com and SimplyHired.com) during July 2008 to April 2009 and extract any text data mentioned about skills in the degree of Computer Science (CS), Management Information Systems (MIS), Computer Information Systems (CIS), and other programs related to computer. The author gets 244,460 jobs as the results. Next, the skills are identified by using tokenization technique to eliminate the general terms, the advertisements that mention less or too much skills, and the skills that have less requirements. The authors use cluster analysis which is a kind of statistic technique for grouping and get 20 clusters which are verified with 100 jobs from random selection. The results from 20 clusters show that IT Managers is the most required position and Security Specialists and Project Analysts/Managers, respectively. The position related to Microsoft technology is 14% and Open Source and Java technology is 12%. The position are grouped again into five clusters which are Web Developers, Software Developers, Database Developers, Managers and Project Analysts (Fig. 3). The results show that Web Development is one fourth to Java Programming. Open Source Development is required highly to 40%. The position of Database Administrator for Oracle is required quite high to 91%. The requirement of these skills and positions analysed from the job advertisements is useful for IT-related workforce and Human Resource (HR) department. However, this study has some limitations as follows: (1) the job advertisement from year 2006 were mostly online, the content repeat as well in the printed advertisements, but the online job advertisements can carry more job details. Therefore, the authors then did not analyse the printed advertisements for this research, and (2) the analysis has done for a period of time, it does not predict the trend and the popular skills for the future requirements.

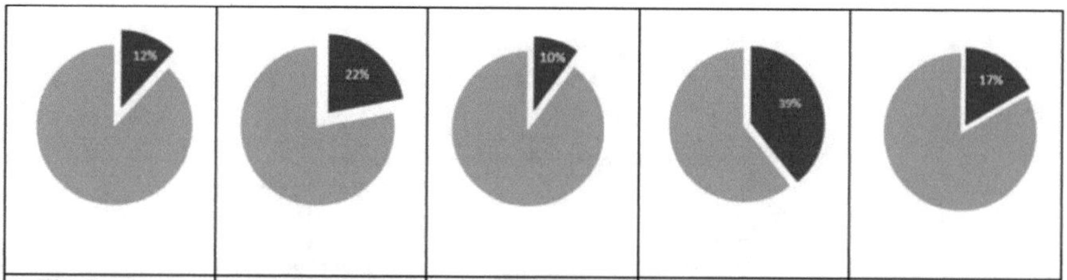

Web Developers	Software Developers	Database Developers	Managers	Project Analysts
General	**General**	**General**	**General**	**General**
■ Web programmers ■ MS web developers	■ Java programmers ■ MS web developers ■ Open source developers	■ Database developers	■ IT managers	■ Project analysts managers
Web Application Developers	**C/C++ Programmers**	**Application Developers**	**Technology**	
■ MS web application project analysts ■ Java database web developers ■ Open source web application developers	■ C/C++ programmers ■ System-level C/C++ programmers	■ Java database application developers ■ MS Visual Basic database application ■ MS database application developers	■ System Administrators ■ Network Administrators ■ Database Administrators ■ Security specialists	

Fig. 3. The clusters of job positions and skills in IT professional.

Debortoli et al. [13] use text mining techniques to analyse job advertisements from the website named www.monster.com. There are 1,357 jobs related to Business Intelligence (BI) and 450 jobs related to Big Data (BD). The jobs are classified and categorized based on the requirements for BI and BD professionals. The research results showed that (1) Information Technology (IT) and business knowledge are two of the major important skills for the achievement of BI and BD projects; (2) BI skills focus on the usage of commercial products from vendor, however, BD skills focus on the software development and statistic knowledge; (3) The requirements for BI professionals in the market is still more than BD professionals, and (4) BD projects need more human-capital-intensive than BI projects. The limitations of the research are as follows: (1) the advertisements were collected between September 2013 and March 2014; (2) the advertisements may not represent the real requirements, and (3) the advertisements selected were only in the English language, however, other languages will be analysed in the future and the research methods to be conducted adjusted to get more proper results.

The results from exploratory data analysis can be used to generate the skills taxonomy as shown in Fig. 4. The requirements in BI skills can be classified into two aspects: Information Technology (IT) and business. Business aspects can be classified into two groups: business domain (e.g., healthcare, digital marketing) and management (e.g., sales and business development, project management). IT aspects can be classified into two groups: concepts and methods (e.g., data administration, software engineering) and products (e.g., BI platform from Microsoft, SAP, SAS, or IBM and Microsoft web portals). The three BI skills which are the most desired from the market are BI platform, healthcare and sales, and business development.

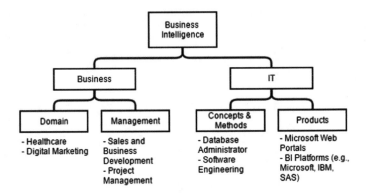

Fig. 4. The taxonomy of BI skills (Source: Debortoli et al. 2014, p. 295)

Marjanovic and Dinter [2] performed semantic text mining with a software named Leximancer for content analysis. The objective was to analyse the BI and Business Analytics-related research publications submitted to The 50th Hawaii International Conference on System Sciences (HICSS) between 1990 and 2016.

The authors stated that BI and Business Analytics (BI&A) have several definitions depending on the industries. Some industries define BI and BA with the same definitions, some industries define differently such as BI refers to the reporting tools and is used as the basics and technical infrastructure, whilst BA is advanced analytics (e.g., predictive analysis, data mining). However, the author define BI and BA to cover the organisations applications infrastructure and procedure.

The two steps for the research methods are as follows: Firstly, to collect articles from the HICSS conference website and verify them with an electronic database such as IEEE Explorer, DBWorld and Google Scholar. There are 140 articles which are collected articles after peer-reviewing and are not older than the year 1990. Secondly, the data are classified into four groups based on the content. Two analyses are performed: (1) descriptive analysis is used to understand the whole picture of the research related to BI and BA using citation analysis, and (2) lexical analysis is used to understand the changes of the research related to BI and BA. All articles are uploaded to Leximancer and follow the data mining process. First, the articles from each group are analysed to see the concept used repeatedly. Next, the concept maps are created for each group. Lastly, if there is similarity among them, they are grouped together.

The descriptive analysis results show that (1) most articles are from three countries (i.e., USA, Australia, and Germany); (2) Winter R. from University of St. Gallen is the author who submitted the most with eight articles; (3) most articles submitted are from University of St. Gallen, Switzerland: 12 articles, and (4) the article titled *"Knowledge Management Systems: Emerging Views and Practices from the Field"* from year 1999 is the article which has the highest citation: 368 times. The results from content analysis shows that between 1990 and 1996 there are more articles related to business and organizational management and Executive Information System (EIS) application; during the period 1997 2003 there are more articles related to data warehouse and performance management; between 2004 and 2011 there are more articles related to Business Intelligence (BI) and Business Analytics (BA); during the period 2012–2016 there are still more articles related to BI and BA however the last two years (2015–2016) more articles about big data are submitted.

The author states that typically the systematic literature review, citation analysis and thematic analysis have been used for the research projects, however, this research uses the text mining concept with Leximancer to enhance the scope of the literature review. Moreover, Leximancer as a text mining tool can visualise the structure of the concepts and contents and be suitable for other analytics such as text mining, qualitative analysis, quantitative analysis, etc.

3   Applied Techniques

The main techniques that will be applied to this research are briefly described and shown in Fig. 5 as follows:

(1) **Job Advertisements Extraction**
About 1,000 BI jobs will be scraped and extracted from the English popular job web portals for ASEAN countries. The same positions will be verified and eliminated. Text data collection will be stored in a table including job position, company name, required skills, offered salary, posted date and working location. The extraction will be performed every two weeks.

(2) **Tokenisation**
The job data collection will be reviewed and preprocessed such as adding attributes (e.g., province and country), separating Thai words and English words and changing capital letters to lower case letters, etc. The dictionary for tokenisation is generated containing keywords related to job positions and jobs/skills description such as analyst, consultant, manager, qlikview, scala, sql, english, computer, science, marketing, warehousing, statistics, logistics, etc. Then the job positions and job/skills description are converted to the document-term matrix using binary term occurrences schema.

(3) **Association Rules**
The relationship between job positions and other information are analysed by using data mining techniques. Association rules are applied to the document-term matrix with FP-Growth method to calculate the frequent itemsets which appear together in the document-term matrix. Later, association rules are created based on the set of frequent itemsets. The patterns of job positions, required skills and other attributes can be discovered.

(4) **Visualisation**
The results from text mining and data mining processes can be visualised using various tools (e.g., RapidMiner) and in several ways (e.g., word cloud, geography and time series).

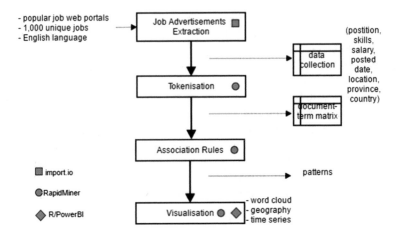

Fig. 5. The main techniques using for job analysis.

4 Preliminary Results

Preliminarily, the popular thirteen job web portals of Thailand were selected and filtered for the job position related to Business Intelligence. Only six of them were able to be extracted by a web scraper import.io and the results were stored in tables. Each web portal has a different layout therefore the data collected from extraction are different. Job information from some job portals can be extracted almost completely, but some can only be extracted partially. However, the information needed for this research is revised manually.

IInitially, 21 job positions were extracted from the www.thJobsDB.com for proving the concept. The capital letters were changed to lower case letters using a text editor Sublime Text, Thai and English information were separated in different columns, non-BI related job positions were removed, and attributes about province and country were added. The job positions and job description were reviewed and the dictionary for tokenisation was created. Next, job information and dictionary were read in order to create the document-term matrix and association rules technique was applied using FP-Growth to calculate the frequent itemsets, generate the set of association rules and represent the knowledge patterns. The results of FP-Growth showed the size of frequent itemsets is between 1 item to 7 items. Figure 6 is a part of the results for the size of 3 items including business intelligence and another word.

The first testing results showed that more job information is needed to see more precise patterns and knowledge, the job information collected from the job portals needed to be corrected and preprocessed, the other extraction tools needed to evaluate job extraction should be performed weekly to see the changes in job posting during time, the creation of the dictionary needed to be improved to support N-gram sequences concept, and the whole process needed to be performed repeatedly to refine better knowledge patterns.

Size	Support	Item1	Item2	Item3
3	0.368	intelligence	business	analyst
3	0.368	intelligence	business	data
3	0.368	intelligence	business	experience
3	0.263	intelligence	business	bi
3	0.263	intelligence	business	sql
3	0.211	intelligence	business	computer
3	0.211	intelligence	business	english
3	0.211	intelligence	business	good
3	0.211	intelligence	business	strong

Fig. 6. A part of the results of calculating the frequent itemsets using FP-Growth 3 items.

References

1. Chiang, R.H., Goes, P., Stohr, E.A.: Business intelligence and analytics education, and program development: a unique opportunity for the information systems discipline. ACM Trans. Manage. Inf. Syst. (TMIS) **3**, 12 (2012)
2. Marjanovic, O., Dinter, B.: 25+ Years of Business Intelligence and Analytics Minitrack at HICSS: a text mining analysis. In: Proceedings of the 50th Hawaii International Conference on System Sciences, pp. 5338–5347 (2017)
3. Wixom, B., Ariyachandra, T., Douglas, D., Goul, M., Gupta, M., Gupta, B., Iyer, L., Kulkarni, U., Mooney, J.G., Phillips-Wren, G., Turetken, O.: The current state of business intelligence in academia: the arrival of big data. Commun. Assoc. Inf. Syst. **34**, 1 (2014)
4. Luhn, H.P.: A business intelligence system. IBM J. Res. Dev. **2**, 314–319 (1958)
5. Grimes, S.: BI at 50 Turns Back to the Future. InformationWeek (2008). http://www.informationweek.com/software/information-management/bi-at50-turns-back-to-the-future/d/d-id/1073576?. Accessed 28 Sept 2015
6. Power, D.J.: A brief history of decision support systems (2007). DSSResources.COM. http://www.groupdecisionroom.nl/artikelen/decisionsupport-system.pdf. Accessed 10 July 2015
7. Elena, C.: Business intelligence. J. Knowl. Manag. Econ. Inf. Technol. **1**, 1–12 (2011)
8. Wixom, B., Watson, H.: The BI-based organization. Int. J. Bus. Intell. Res. **1**, 13–28 (2010)
9. Chiang, R.H.L., Goes, P., Stohr, E.A.: Business intelligence and analytics education, and program development. ACM Trans. Manag. Inf. Syst. **3**, 12:1–12:13 (2012)
10. Tan, A.-H.: Text mining: the state of the art and the challenges. In: Proceedingsof the PAKDD 1999 Workshop on Knowledge Discovery from Advanced Databases, vol. 8, pp. 65–70 (1999)
11. Cai, Y., Sun, J.-T.: Text Mining. Encyclopedia of Database Systems, pp. 3061–3065. Springer (2009)
12. Litecky, C., Aken, A., Ahmad, A., Nelson, H.J.: Mining for computing jobs. Comm. ACM. **48**, 9194 (2005)
13. Debortoli, S., Muller, O., vom Brocke, J.: Comparing business intelligence and BigData Skills. Bus. Inf. Syst. Eng. **6**, 289–300 (2014)

The Model of Teenager's Internet Usage Behavior Analysis Using Data Mining

Thidarat Pinthong[(⊠)] and Worawut Yimyam

Department of Computer Business, Phetchaburi Rajabhat University,
Pho Rai Wan, Thailand
thidarat.pin@mail.pbru.ac.th,
worawut_yimyam@hotmail.com

Abstract. This researcher aimed to study teenagers' Internet usage behavior and to investigate associations between parental raising and teenagers' Internet usage behavior and experience. Data mining was employed to cluster parents and teenagers in order to identify associations. The results showed that parent clusters had no correlation with teenager clusters and teenager clusters and parent clusters were less correlated when compared to parent clusters themselves. Therefore, teenagers' Internet usage behavior was not correlated to parental raising.

Keywords: Data mining · Grouping · Relationship

1 Introduction

Nowadays, Internet technology enables people to communicate worldwide [1, 2] through advanced devices such as personal computers, laptops, mobile phones and tablets etc. With the potential of Internet, many services are available including websites, emails, blogs, chats etc. As a result, there are many users using these services. The most frequently Internet using group is teenagers aged between 10–25 years old [3] and it tends to increase more and more. Teenagers' Internet activities include social networking, emails and online games etc. Internet services are widely available. They have both beneficial usage such as social networking or emails for educational purposes etc. and non-beneficial usage such as pornography websites or violent online games etc. Teenagers are exposed to the risk of accessing wrong data. Consequently, it is essential to control and provide advice on appropriate Internet usage to them.

This research aimed to study teenagers' Internet usage behavior using data mining technique to cluster teenagers and to investigate associations between parental raising and teenagers' Internet usage behavior and experience.

The content of the study is divided as follows; Sect. 2: Theories and Related Research; Sect. 3: Research Methodology; Sect. 4: Results and; Sect. 5: Summary and Suggestions.

© Springer Nature Switzerland AG 2019
T. Theeramunkong et al. (Eds.): iSAI-NLP 2017, AISC 807, pp. 196–203, 2019.
https://doi.org/10.1007/978-3-319-94703-7_18

2   Theories and Related Research

There are many studies about analysis of teenagers' Internet usage behavior using data mining techniques. An analysis of teenagers' Internet usage behavior is very important for monitoring teenagers' Internet usage and for providing appropriate advice on Internet usage to teenagers. The most frequently Internet using group is females aged between 15–25 years old [1]. Previous study investigated 690 teenage students and found that the most frequently used Internet was Social Media. Many research studies applied Association Rules to identify associations. These include a study of navigational behavior to predict web transactions [2], a study of data grouping using Cluster technique, a study of Self-Organizing Maps [3] employed to group Internet usage on WLAN, DBANSCAN and K-mean techniques for grouping traffic on Internets [4] and Hierarchical Cluster technique for grouping [5] etc.

2.1   Cluster

Cluster is one of data mining technique that considers Similarity and Dissimilarity or Distance of data which is unsupervised learning. There are many cluster techniques as follows.

2.2   DBScan (Density-Based Spatial Clustering)

DBScan [4, 6] is density-based clustering using a distance from data by identifying Epsilon and the least number of data within the Epsilon (MinPoints). This technique is appropriate for huge amount of data containing lots of Noise. Euclidean Distance is measured.

2.3   K–Means Cluster

K-Means [4, 7] technique applies Partition grouping. K value is identified. An algorithm randomly selects K number which is Centroids value or an average value. Then, a distance between data and Centroids value is measured in order to group data and identify an average value and a distance of data is measured until the average data is stable.

2.4   Hierarchical Cluster

Hierarchical [5] technique is a cluster technique using hierarchical concept which is similar to a tree. It calculates similarity and distance of data in forms of similarity and distance matrix.

2.5   Self-Organizing Maps Cluster

Self-Organizing Maps [3] is a technique that groups data through maps or grids. It performs as neuron network with unsupervised learning consisting of input vector set. Output consists of many neurons. Learning of SOM is adjusting weights affecting input.

2.6   Association Rules

Association Rules [2] is one of data mining used to identify relationships between 2 sets of data or more. In order to identify association rules, support value and confidence value must be calculated. All values are compared with identified minsup and minconf values so as to reduce data association.

3   Research Methodology

In this study, the researcher applied clustering technique and identified data association in order to investigate teenagers' Internet usage behavior. The research methodology was designed as follows.

3.1   Data Preparation

The researcher collected data from 800 questionnaires on teenagers' Internet usage. The target groups were African-American families and Latin families. They were divided into 2 groups including parent group and teenager group. The collected data was processed as follows.

3.1.1 Data Cleaning is a process to filter 800 sets of data. Only 748 sets of data were left.

3.1.2 Data Select

After filtering the data, the data was separated into 2 types: data of parents and data of teenagers. Tables 1 and 2 illustrate the selected data.

Table 1. Data of parents

No	Name	Detail	Type
1	Psex	Genders of parents	Input
2	Page	Ages of parents	Input
3	Pfriend	Being friends with teenagers on Social Network	Input
4	Pcheckdata	Data on teenager monitoring	Input
5	Padvice1	Advice on safe Internet usage for teenagers	Input
6	Padvice2	Advice on having good behavior towards others on online world	Input
7	Ptalk	Talks about actions on Internet	Input
8	Ptalkshare	Talks about to-share and not-to-share issues on online world	Input
9	Pcontrol	Control of teenagers' online activities	Input
10	Pcheckweb	Monitor of teenagers' websites	Input
11	Plimit	Restriction of teenagers' mobile phone usage	Input

3.1.3 Data Transformation

It is a process of transforming selected data so as to group the data and identify the association.

3.2   Cluster Data

The researcher applied 4 types of data clustering as shown in Table 3. The data cluster was divided into 2 groups: (1) parental raising clusters and; (2) teenagers' Internet usage behavior and experience clusters.

Table 2. Data of teenagers

No	Name	Detail	Type
1	Tsex	Genders of teenagers	Input
2	Tage	Ages of teenagers	Input
3	Tusenet	Internet usage frequency	Input
4	Tgame	Internet usage and game consoles	Input
5	Tmp3	Internet usage and MP3	Input
6	Ttablet	Internet usage and Tablet	Input
7	Ttwitter	Internet usage on Twitter	Input
8	Tskype	Internet usage on Skype	Input
9	TBpost	Decision on not expressing opinions causing negative effects to themselves in the future	Input
10	TBstop	Frequency of giving advice to others to stop inappropriate action on online world	Input
11	TBprotect	Frequency of protecting victims threatened on online world	Input
12	TBbjoin	Frequency of taking part in threatening others on online world	Input
13	tbignore	Frequency of ignoring threatening situations on online world	Input

Table 3. Details of experimental models

Model	Name	Abbreviation
1	Density-based spatial clustering	DBScan
2	K–means	KM
3	Hierarchical clustering	HC
4	Self-organizing maps	SOM

3.3   Identifying Efficiency and Selecting Clustering Format

The researcher identified efficiency of each cluster and selected the most efficient cluster in order to find data associations using the following equation.

$$RMSSTD = \sqrt{\frac{\sum_{i=1}^{k} \sum_{j=1}^{p} \left(x_{ij} - \bar{x}_{ij}\right)^2}{\sum_{i=1}^{k} \sum_{j=1}^{p} \left(n_{ij} - 1\right)}} \tag{1}$$

When k is number of clusters, p is independent variables within set of data, x_ij is i set and j independent variable, x ̄_ij is an average value of i set and j independent variable and n_ij is number of data in i set and j independent variable.

3.4   Identifying Data Association

The researcher identified associations of collected data into 4 types: (1) associations between parental raising and teenagers' Internet usage behavior based on parental raising cluster; (2) associations between parental raising and teenagers' Internet usage experience based on parental raising cluster; (3) associations between parental association and teenagers' Internet u6sage behavior based on Internet usage behavior cluster and; (4) associations between parental raising and teenagers' Internet usage experience based on Internet usage experience cluster.

4   Experimental Results

The research conducted the research and summarized the results into 3 groups as follows.

4.1   Results of Identifying Efficiency

Identifying efficiency of clusters by selecting the most efficient cluster was divided into 2 types as follows.

4.1.1   Results of Identifying Efficiency of Parent Clusters

Identifying efficiency of parent clusters so as to select the most efficient cluster is shown in Table 4.

Table 4. Efficiency values of parent clusters

Cluster	DBSCAN	K-means	Hierarchical	SOM
2	–	0.4223	–	0.4693
3	–	0.4392	–	0.4590
4	–	0.3806	–	0.4590
5	–	0.3943	–	0.4655

Based on the table, the most efficient cluster is K-means that had 4 groups. The value is 0.3806 which is the minimum value. DBSCAN and Hierarchical cannot measure efficiency because DBSCAN yielded only one cluster and Hierarchical cluster was too widely distributed.

4.1.2   Results of Identifying Efficiency of Teenager Clusters

The most efficient parent clusters was applied to teenager clusters. The clusters were divided into 2 types based on: (1) Internet usage behavior and; (2) Internet usage experience as shown in Tables 5 and 6.

According to Tables 5 and 6, the most efficient clusters include 5 clusters with the values of 0.5033 and 0.4108 respectively.

Table 5. Efficiency values of teenager clusters base on internet usage behavior

Behavior	2 Clusters	3 Clusters	4 Clusters	5 Clusters
1	0.5342	0.4994	0.5029	0.5014
2	0.5011	0.5310	0.4990	0.5010
3	–	0.5066	0.5139	0.4998
4	–	–	0.5066	0.5103
5	–	–	–	0.5040
Average	0.5176	0.5123	0.5056	0.5033

Table 6. Efficiency values of teenager clusters based on internet usage experience

Experience	2 Clusters	3 Clusters	4 Clusters	5 Clusters
1	0.5463	0.5126	0.3555	0.4444
2	0.5545	0.3593	0.5065	0.4799
3	–	0.5513	0.5366	0.5319
4	–	–	0.4655	0.3160
5	–	–	–	0.2820
Average	0.5504	0.4744	0.4660	0.4108

4.2 Results of Clusters

Results of clusters are divided into 2 types as follows.

4.2.1 Results of Parent Clusters

Efficiency values of parent clusters were tested, measured and divided into 4 clusters as shown in Table 7.

Table 7. Parent clusters

Cluster 1	Female, talk about what-to-share matters with teenagers
Cluster 2	Female, provide advice on good behavior toward others
Cluster 3	Female, provide advice on safe Internet usage
Cluster 4	Talk about actions on Internet

4.2.2 Results of Teenager Clusters Based on Internet Usage Behavior and Experience

Efficiency values of teenager clusters were tested, measured and divided into 5 clusters as shown in Tables 8 and 9.

Table 8. Results of teenage clusters based on internet usage behavior

Cluster 1	Age < 14 years
Cluster 2	Female, age > 14 years, use Facebook
Cluster 3	Male, age > 14 years, not use Facebook
Cluster 4	Age > 14 years, use Facebook, use Tablet
Cluster 5	Female, age > 14 years, use Facebook, use Tablet

Table 9. Results of teenage clusters based on internet usage experience

Cluster 1	Never threaten others
Cluster 2	Female, age > 14 years, never threaten others
Cluster 3	Male, age > 14 years, never threaten others
Cluster 4	Male, stop others from threatening others sometimes
Cluster 5	Female, stop others from threatening others sometimes

4.3   Results of Identifying Associations Between Parental Raising and Teenagers' Internet Usage Behavior and Experience

Identifying associations between parental raising and teenagers' Internet usage behavior and experience yielded many association rules which can be divided into 4 types as follows.

4.3.1 Associations between parental raising and teenagers' Internet usage behavior based on parental raising clusters yielded 4 association rules as follows. (1) Teenagers never been using Twitter would not be friends with their parents on social network but their Internet usage was controlled by parents. (2) Teenagers whose parents never controlled and monitored their website usage had used Facebook. (3) Internet usage of teenagers using Facebook had never been controlled. (4) Teenagers using Facebook had been friends with their parents on social network.

4.3.2 Associations between parental raising and teenagers' Internet usage experience based on parental raising clusters showed that parental raising had no correlation with teenagers' Internet usage experience.

4.3.3 Associations between parental raising and teenagers' Internet usage behavior based on teenagers' Internet usage behavior yielded 5 association rules as follows. (1) Teenagers never been using Twitter had received advice on safe Internet usage from their parents. (2) Teenagers whose parents provided advice on safe Internet usage to them had used Facebook. (3) Teenagers whose parents talked about what-to-share and what-not-to-share issues to them had used Facebook. (4) Teenagers whose parents talked about what-to-share and what-not-to-share issues and provided advice on how to behave well to others to them were older than 14 years old. (5) Teenagers whose parents talked about actions on Internet had used Facebook.

4.3.4 Associations between parental raising and teenagers' Internet usage experience based on Internet usage experience yielded 5 association rules as follows. (1) Parents would talk about what-to-share and what-not-to-share issues with teenagers aged lower than 14 years. (2) Teenagers deciding not to post things giving negative effects to themselves had talked to parents about what-to-share and what-not-to-share issues. (3) Teenagers whose parents provided advice on how to behave themselves well to others on online world had never threatened others. (4) Teenagers whose parents talked about what-to-share and what-not-to-share issues and who had never threatened others were older than 14. (5) Teenagers who had never threatened others talked to their parents about what-to-share and what-not-to-share issues.

5 Conclusion

The analysis of teenager's Internet usage behavior using data mining technique to cluster groups of parents and teenagers based on teenagers' Internet usage behavior and experience in order to identify associations of parental raising and teenagers can be summarized as follows. Parent clusters had no correlation with teenager clusters. However, there was the correlation between teenager clusters and parent clusters. That is, association rules occurring within teenager and parent clusters were less than those within parent clusters themselves. Therefore, teenagers' Internet usage behavior had no correlation with parental raising. Teenagers can access Internet because plenty of Internet devices are available and inexpensive. As a result, it is difficult to control and monitor.

This research applied questionnaires which may require more factors about parents affecting teenagers' behavior. For instance, factors about social network usage should be added. In further studies, the researcher will study how parents' social network usage behavior is correlated with teenagers' social network usage. It can be applied to plan, control and monitor Internet usage appropriately.

References

1. Lin, Y.H., Liang, J.C.: Preschool teachers' internet attitude and their internet selfefficacy: a comparative study between pre-service and in-service teachers in Taiwan. In: 2012 IIAI International Conference on Advanced Applied Informatics (IIAIAAI), pp. 334–339. IEEE (2012)
2. Ling, C.S., et al.: Malaysian internet surfing addiction (MISA): factors affecting the internet use and its consequences. In: 2011 IEEE International Conference on Computer Applications and Industrial Electronics (ICCAIE). IEEE (2011)
3. Liu, Q.-X., et al.: Parent–adolescent communication, parental Internet use and Internet specific norms and pathological Internet use among Chinese adolescents. Comput. Hum. Behav. **28**(4), 1269–1275 (2012)
4. Ketcham, M., Ganokratanaa, T., Srinhichaarnun, S.: The intruder detection system for rapid transit using CCTV surveillance based on histogram shapes. In: 2014 11th International Joint Conference on Computer Science and Software Engineering (JCSSE), pp. 1–6. IEEE (2014)
5. Ketcham, M., Yimyam, W., Chumuang, N.: Segmentation of overlapping Isan Dhamma character on palm leaf manuscript's with neural network. In: Recent Advances in Information and Communication Technology, pp. 55–65. Springer (2016)
6. Khakham, P., Chumuang, N., Ketcham, M., Dhamma, I.: Handwritten characters recognition system by using functional trees classifier. In: 11th International Conference on Signal-Image Technology & Internet-Based Systems (SITIS), Bangkok, pp. 606–612 (2015). https://doi.org/10.1109/sitis.2015.68
7. Yimyam, W., Ketcham, M.: The automated parking fee calculation using license plate recognition system. In: International Conference on Digital Arts, Media and Technology (ICDAMT), pp. 325–329. IEEE

Information Security and Privacy (ISA)

Privacy-Preserving Reputation Management in Fully Decentralized Systems: Challenges and Opportunities

Ngoc Hong Tran[1(✉)], Leila Bahri[2], and Binh Quoc Nguyen[1]

[1] Vietnamese German University, Thủ Dầu Một, Vietnam
{ngoc.th,binh.nq}@vgu.edu.vn
[2] Koc University, Istanbul, Turkey
lbahri@ku.edu.tr

Abstract. Reputation is one of crucial personal information strongly attached to each person, so it affects directly to its owner in whatever way it is used. Any form of reputation violation can carry a serious consequence to its owner. Therefore, reputation needs to be managed in a secure way. Moreover, reputation management is a substantial process as it plays a key role in building up a certain trust level, among just-met users. Thus just-met users can base on reputation scores to make a decision on starting up their communication. In addition, reputation management reduces risks of leaking user privacy and losing data security. So far, several works on managing reputation in online social network have been studied, mostly using the Internet for data transmission. In this work, reputation management is placed into the context of fully decentralized environment. Under the more harsh conditions which the decentralized mobile environment carries, such as no central node, frequently changed user location, and restrictions of energy, power, memory, etc., and the security requirement, methods of reputation management get more challenged. Various security problems of managing reputation are still open to be solved. In this work, we make a discussion about recent works on privacy-preserving reputation management in a decentralized environment. We then present challenges, and conclude the open problems as well as possible solutions in reputation management.

Keywords: Reputation management · Identity validation
Decentralized model · Mobile environment · Peer-to-peer (P2P) network
Privacy preserving · Security · Encryption · Anonymity

1 Introduction

Thus far, practical systems have mainly evolved around logically centralized models for the management of identity, data, wealth and computing [RAV16]. Centralization requires trust in a service provider, but the continuously reported incidents of surveillance and privacy breaches (e.g., the Snowden affair [OHA14]) have shown that this trust can be broken. In addition to that, centralization creates single points of failure, that once they become out of service, no one can benefit from them anymore.

© Springer Nature Switzerland AG 2019
T. Theeramunkong et al. (Eds.): iSAI-NLP 2017, AISC 807, pp. 207–215, 2019.
https://doi.org/10.1007/978-3-319-94703-7_19

Centralization also requires huge investments (e.g., in building data centers), that once established by one party create obstacles of entry to competitors, creating as such a service market dominated by monopoles (e.g., Facebook, Google, etc.). These issues have been calling into question the centralized model for computing, and researchers have reacted by giving a considerable focus on the development of decentralized peer-to-peer (P2P) computing. In P2P computing, services are supposed to be provided by the collaboration of various free peers (i.e., nodes), that put their computing power into the network and contribute in turning the service's protocol. Although such a setting seems to be providing more scalability in terms of both performance and service provision, more privacy, and a more flexible business model within which market entry is more affordable, it does also present tremendous challenges. One of these are related to the security of the P2P network, in terms of the honesty of the participating nodes. One of the main techniques around this problem is reputation management.

Reputation management is the process by which entities in the system are assigned reputation scores that increase or decrease based on their observed behavior. Reputation management is a required in both centralized and decentralized systems, but with mostly different logical needs. In fact, in centralized systems, reputation management is a process carried out by an already trusted service provider, and reputation scores are assigned to end-users based on their centrally observed behavior when using the centrally provided service. That is, the central provider has the full view of usage behaviors of all users, and can impose a corresponding technique to compute users reputation. The trusted central provider makes it straightforward for all other users to trust in the corresponding assigned reputation scores to themselves and to the other users.

However, in decentralized system, reputation management becomes a required process not only for end-users but also for the nodes making part of the underlying service provision P2P network. Moreover, in decentralized environments, there is no central trusted authority that observes all behaviors and that is trusted to certify both the mechanism adopted and the reputation scores assigned. These differences, makes it technically challenging to devise a reputation management system in a decentralized setting. In this survey position paper, we focus on the main challenge that the insurance of the security of the decentralized reputation system, in the sense that the assigned scores can be certified, and users or peers cannot tamper with scores, including their own.

This paper is organized as follows: Sect. 1 provides a basic introduction about reputation management and motivation of this work as well as the paper structure. In Sect. 2, we survey the recent works on decentralized reputation management. Then we make a discussion about challenges we face nowadays for a reliable reputation management, particularly in the peer to peer environment with many rough constraints in Sect. 3. Then Sect. 4 presents opportunities and possible research directions based on the mentioned challenges. We make a conclusion of this work in Sect. 5.

2   Decentralized Protocols for Reputation Management

Reputation score can be obtained by doing evaluation by the surrounding users. For example, trading and sharing systems such as e-commerce, goods selling, hotel booking, ridesharing, etc. where agents do not know each others, it seems hard for the involver to make the right choice. They, therefore, should rely on the feedback about services or goods they intent to choose. Provider and customer can rate each other after trading or sharing. Peers can learn about the reputation of the others to make a good decision. [JOS16] points out that reputation allows agents to build trust and to make agents accountable for their behavior.

However, for reputation's importance, attackers pay much attention to find a way abusing the other reputation for their further target. As in [HOF09], authors mentioned several rational attacks which can be made when the agents are benefit by deviating from them, including self-promotion, whitewashing, slandering, and denial of service. Hence, they need a tool protect their privacy.

Technically, whenever user privacy is in need, anonymity techniques are considered as the first solution to be used. It can cover the real user ID by a coding layer obfuscating the observance of attackers. As in the works [ELI08, PET], authors built up pseudonym based schemes which require a Trusted Third Party (TTP). They therefore cannot be considered truly decentralized. Additionally, Anceaume et al. [ANC12] proposed a decentralized anonymity-preserving reputation system prevents Sybil attacks by charging a fee. In Lajoie Mazenc et al.'s solution [LAJ15], accredited signers are required to make resource heavy calculations for each rating of each Service Provider. Soska et al. [SOS16] proposed a decentralized anonymous marketplace called Beaver that is resistant against Sybil attacks on vendor reputation, while preserving user anonymity. By employing some cryptographic primitives such as non-interactive zero-knowledge proofs and linkable ring signatures, Beaver attain high levels of usability and practicality, and strong anonymity for buyer. However, it only offers limited protection for vendors and reveals many other weaknesses such as the problems of unclaimed reviewing fees, collusion attacks, advanced searching, etc. Bazin et al. [BAZ16], proposed a decentralized anonymity-preserving reputation system. In this system, the feedback will be sent by the client after the transaction a changable timeout. Merkle trees and signed blocks of data are used to minimize the workload on the trackers and to fairly distribute the record maintenance tasks to the service providers. It is proved to be efficient, anonymity-preserving, decentralized, and robust against various known attacks against reputation systems, such as ballot-stuffing and Sybil attacks.

Indeed, anonymity techniques have first been designed for static collections of data; however, with the emergence of situations and scenarios requiring the anonymization of dynamic and real time data (such as online data-streams, geolocalization information, etc.), these techniques have been revised to fit such scenarios as well [BAR08, GUO13]. Works such as [MAS11] or [LIO13] have already considered anonymization techniques via generalizations for real time geo-localization data. Therefore, a desired protocol should be satisfied the criteria of anonymity, low overhead, and proper management of new agents.

There are also some approaches on data hiding or cryptography. In another work [PAV04], Pavlov et al. proposed one of the first privacy preserving reputation systems. They protect confidentiality of feedback by hiding the secure sum and verifiable secret sharing of the submitted ratings. In the meanwhile, approaching another methods based on cryptographic algorithms, Dolev et al. [DOL14] can only make a conditionally secure system using homomorphic and commutative cryptography. However, it cannot be resistant against large numbers of colluding malicious adversaries. As well, Hasan et al. [HAS12, HAS13] as well as Dimitriou et al. [DIM14] introduced some systems based on additive homomorphic cryptography and Zero-Knowledge proofs where privacy of a given user can be preserved even in the presence of a large majority of malicious users. However, these protocols cannot hide the list of users who participated in the rating.

Moreover, we can find some other techniques can be used for reputation management. Based on a blockchain, Schaub et al. [SCH16] secured ballot-stuffing. Bethencourt et al.' protocol used signatures on published data. However, attacker may take advantage of his old good reputation without being affected by any new dissatisfaction that his recent activity because the feedback is monotonic.

For applications, another well-known protocol for this problem is EigenTrust [JOS16] which meet most of the requirements. This co-utile protocol can be applied to a variety of scenarios and heterogeneous reputation needs. Some extended version of it [KAM03] tried to provide a more secure distributed calculation that cancels benefits of deviating from the protocol. Sanchez et al. [SAN16] focused on privacy concerns and the lack of trust of ridesharing. They proposed a fully decentralized P2P ridesharing management network that brings trust among peers. This is based on the co-utility idea [DOM15, DOM16], like invisible hand theory, which makes rational (even selfish) user collaborate and follows the proposed protocols in a self-enforcing way because of their own benefit. This benefit also directs them out of attack these system.

3   Challenges

As mentioned in Sect. 2, mobile user's reputation in the decentralized environment gets more delicate and easier to be defrauded and violated, since there is no any central system managing users as well as doing authentication and authorization, like in the centralized network. The situation gets more harsh when a peer enters a new geographical area, and interacts with strangers nearby. Trusting wrong user can give the malicious user a chance of misusing user's reputation. This then carries a serious consequence to the reputation owner as the reputation affects directly to the others' trust in them. In addition, restrictions of decentralized mobile environments, such as weak power, weak computing performance, lack of physical resources, etc. also contribute to make the environment. Moreover, reputation computation has its own higher security constraints. Furthermore there are still the other edges essentially taken into consideration as trustworthiness estimation, identity validation, owner authentication, as well as privacy preservation problems. In this section, each of challenges is discussed as follows.

3.1   Identity Validation

Identity validation is one of major concerns relating to reputation management, particularly in online social network (OSN), since an accurate identity validation can support the process of reputation use or user's credentiality evaluation to be performed in a more trustworthy manner. Well validating identity intensifies the security of reputation computing too. Based on such benefits for reputation management, identity has been a very interesting problem taking much effort of researchers. Some recent works investigated to validate user identity in OSN [SIR12, CHA11, ROF13, GOG13]. In some other works, authors focused on another side of identity validation as detecting the fake user's profile on OSN [LIM08, CLO11, THE14, CAQ12], while some others studied how to enhance the environment safety for OSN users.

In [BAH14], the authors proposed an identity validating model, by learning the correlations between attributes of users' profiles so the process can achieve the identity perspective. The model also manipulate the profile's trustworthiness validation. For example, attributes (*Job, Education*) could be used for inferring a trustworthy profile on OSN. The authors exploited a supervised and feedbackbased approach by which a group of trusted users provided feedback on the correlations between attribute values. Then the learned correlations are used for evaluating the trustworthiness of new target profiles. The approach in [BAH14] can be considered as a solution for quantifying trustworthiness of an OSN user's profile and aiming to give a proof for the reputation evaluation process. However, when this solution is applied to the mobile decentralized environment, still many points need to be revised, such as autonomous trustworthiness quantification and authentication.

3.2   Privacy Challenges in Autonomous Identity Validation

In decentralization model, identity validation is not performed by any central node. It should be done independently at each peer in the network. The workload from the server now is divided to peers. Peers need to authenticate, authorize, data synchronization, and data coordination between them with their communication partners. The interaction only between peers can carry various risks as one peer itself cannot learn the necessary information about its partners in the network. Therefore, the identity validation in the decentralization model make more challenges to peers than the centralized model [FUR13]. Any information revealed not to the trustworthy users can be abused for further hazardous goals. Privacy can be breached as a result of personal data inference from provided information, or leaked while personal information is transmitted over the underlying network.

Although anonymization techniques have largely used for privacy preservation across a diversity of scenarios and application domains, they do not ensure complete guarantees on privacy. Moreover, these techniques have been proved to be vulnerable to some analysis attacks especially when attackers can access to some additional information that is not then saved into the targeted dataset [CLI13]. In the other hand, these reputation data often include the users' private data such as their name, location, habits, etc. Although there are some syntactic anonymization techniques [CLI13] such as k-anonymity [SWE02], l-diversity [MAC07], t-closeness [LIO07], β-likeness

[CAJ12], etc. for identity disclosure and personal information inference, this is still a problem where a centralized reputation management can enable the threat of compilation, aggregation, exploitation on these data, or even provide the bias matches. This central point of failure can also become an target for external attacks.

Newer research on privacy suggested the concept of differential privacy. Actually, differential privacy can be achieved by adding noise to database to obfuscate the real content [CLI13]. While some researchers believe that differential privacy is the answer to the new requirements on data privacy, others still find it immature to completely replace the well established syntactic alternatives [CLI13]. We believe that the two approaches are equally important in the sense that each of them answers or fits different requirements and different scenarios.

4   Opportunities and Possible Research Paths

As presented in Sect. 3, the mentioned challenges open further requirements for researchers on the reputation management. In this section, we discuss about abilities of reputation management autonomy and of reputation privacy preservation, as follows:

Autonomous reputation management. Each user in the peer-to-peer network needs to manage their own reputation score in an autonomous way. It means that the involved users have to store their reputation score locally, and have to validate the other partners' reputation by themselves without the support of any the third party or server, before deciding to make any trade with the others. Hence, there is a need of an effective protocol for users in the decentralized systems to store and use the reputation scores. Additionally, there is also a need of evaluating correctly a user based on its over-handed reputation score. Moreover, rating the partner's reputation after performing any trade is requested to be honest and correct as well.

Reputation Privacy Preservation. So as to avoid an incorrect identity validation which may cascade the user reputation degradation, anonymization and pseudonym techniques are first thought about as effective solutions for this problem. However, they still have drawbacks which may cause another analysis attacks. Hence, to improve the reputation management, cryptographic algorithms, e.g., hash function, homomorphic encryption, etc., are considered as the powerful means to strengthen the ability of protecting user personal information against the stealing behaviors of malicious insiders or outsiders, and expected to be strongly against any malicious behavior. Anonymity techniques and cryptographic algorithms are always considered as two competitive solutions, as anonymity costs less time and space than cryptographic algorithms. While, encryption takes more time and memory of the system. However, it is not the truth like that. Combinations of the two can carry the wonderful result in protecting the privacy against adversaries. However, up to now there have not existed any effective solution for all scenarios, particularly, most of systems have very strict constraints in performance and time.

5   Conclusion

We survey the current solutions for the problem of privacy-preserving reputation management in decentralized network model. We also discuss about their own challenges in protecting the reputation score against malicious users, as well as emphasize the currently open problems relating to privacy preservation in reputation management systems.

This work may be the first step for us to continue the future works, including investigating more deeply cryptographic algorithms to serve for preserving the privacy of reputation, then proposing a secure protocol for just-met mobile users in decentralized systems to interact together based on the trustworthiness established from their reputation.

References

[ANC12]   Anceaume, E., Guette, G., Lajoie Mazenc, P., Prigent, N., Tong, V.V.T.: A Privacy Preserving Distributed Reputation Mechanism (2012). https://hal.archivesouvertes.fr/hal-00763212

[BAR08]   Cao, J., Carminati, B., Ferrari, E., Tan, K.L.: CASTLE: a δ-constrained scheme for k_s-anonymizing data streams. In: IEEE 24th International Conference on Data Engineering (ICDE 2008), pp. 1376–1378 (2008)

[BAH14]   Bahri, L., Carminati, B., Ferrari, E.: Community-based identity validation on online social networks. In: IEEE ICDCS, pp. 21–30 (2014)

[BAZ16]   Bazin, R., et al.: A decentralized anonymity-preserving reputation system with constant-time score retrieval. IACR, p. 146 (2016)

[CAQ12]   Cao, Q., Sirivianos, M., Yang, X., Pregueiro, T.: Aiding the detection of fake accounts in large scale social online services. In: Proceedings of the 9th USENIX Conference on Networked Systems Design and Implementation, p. 15 (2012)

[CAJ12]   Cao, J., Karras, P.: Publishing microdata with a robust privacy guarantee. Proc. VLDB Endowment **5**(11), 1388–1399 (2012)

[CHA11]   Chairunnanda, P., Pham, N., Hengartner, U.: Privacy: gone with the typing! identifying web users by their typing patterns. In: IEEE PASSAT (2011)

[CLI13]   Clifton, C., Tassa, T.: On syntactic anonymity and differential privacy. In: ICDE Workshops, pp. 88–93 (2013)

[JIN11]   Jin, L., Takabi, H., Joshi, J.B.D.: Towards active detection of identity clone attacks on online social networks. In: ACM CODASPY (2011)

[DIM14]   Dimitriou, T., Michalas, A.: Multi-party trust computation in decentralized environments in the presence of malicious adversaries. Ad-Hoc Netw. **15**, 53–66 (2014)

[DOL14]   Dolev, S., Gilboa, N., Kopeetsky, M.: Efficient private multiparty computations of trust in the presence of curious and malicious users. J. Trust Manage. **1**, 1–8 (2014)

[DOM15]   Domingo-Ferrer, J., Soria-Comas, J., Ciobotaru, O.: Co-utility: self-enforcing protocols without coordination mechanisms. In: Proceedings of the 5th International Conference on Industrial Engineering and Operations Management-IEOM, pp. 1–7 (2015)

[DOM16]   Domingo-Ferrer, J., Sanchez, D., Soria-Comas, J.: Co-utility: self-enforcing collaborative protocols with mutual help. Prog. Artif. Intell. **5**(2), 105–110 (2016)

[ELI08]   Androulaki, E., Choi, S.G., Bellovin, S.M., Malkin, T.: Reputation systems for anonymous networks. In: Proceedings of the 8th International Symposium on Privacy Enhancing Technologies, PETS. Springer, Heidelberg (2008)

[FUR13]   Furuhata, F., Dessouky, M., Ordoez, F., Brunet, M.E., Wang, X., Koening, S.: Ridesharing: the state-of-the-art and future directions. Transp. Res. Part B **57**, 28–46 (2013)

[GOG13]   Goga, O., Lei, H., Parthasarathi, S.H.K., Friedland, G., Sommer, R., Teixeira, R.: Exploiting innocuous activity for correlating users across sites. In: International World Wide Web Conferences Steering Committee, WWW (2013)

[GUO13]   Guo, K., Zhang, Q.: Fast clustering-based anonymization approaches with time constraints for data streams. Knowl. Based Syst. **46**, 95–108 (2013)

[HS12]   Hasan, O., Brunie, L., Bertino, E.: Preserving privacy of feedback providers in decentralized reputation systems. Comput. Secur. **31**(7), 816–826 (2012)

[HAS13]   Hasan, O., Brunie, L., Bertino, E., Shang, N.: A decentralized privacy preserving reputation protocol for the malicious adversarial model. IEEE Trans. Inf. Forensics Secur. **8**(6), 949–962 (2013)

[HOF09]   Hoffman, K., Zage, D., Nita-Rotaru, C.: A survey of attack and defense techniques for reputation systems. ACM Comput. **42**(1), 1 (2009)

[JOS16]   Domingo-Ferrer, J., Farràs, O., Martínez, S., Sánchez, D., Soria-Comas, J.: Self-enforcing protocols via co-utile reputation management. Inf. Sci. **367**(C), 159–175 (2016). https://doi.org/10.1016/j.ins.2016.05.050

[KAM03]   Kamvar, S.D., Schlosser, M.T., Garcia-Molina, H.: The EigenTrust algorithm for reputation management in P2P networks. In: Proceedings of the 12th International Conference on World Wide Web, pp. 640–651 (2003)

[LAJ15]   Lajoie-Mazenc, P., Anceaume, E., Guette, G., Sirvent, T., Tong, V.V.T.: Efficient distributed privacy-preserving reputation mechanism handling non-monotonic ratings (2015). https://hal.archives-ouvertes.fr/hal-01104837

[LIO07]   Li, N., Li, T., Venkatasubramanian, S.: t-Closeness: privacy beyond k-Anonymity and l-Diversity. ICDE **7**, 106–115 (2007)

[LIO13]   Li, H.P., Hu, H., Xu, J.: Nearby friend alert: location anonymity in mobile geosocial networks. IEEE Pervasive Comput. **12**(4), 62–70 (2013)

[YU008]   Yu, H., Gibbons, P.B., Kaminsky, M., Xiao, F.: Sybillimit: Anear-optimal social network defense against sybil attacks. IEEE Security and Privacy (2008)

[MAC07]   Machanavajjhala, A., Kifer, D., Gehrke, J., Venkitasubramaniam, M.: l-diversity: privacy beyond k-anonymity. ACM Trans. Knowl. Discov. Data (TKDD), **1**, 1–3 (2007)

[MAS11]   Mascetti, S., Freni, D., Bettini, C., Wang, X.S., Jajodia, S.: Privacy in geo-social networks: proximity notification with untrusted service providers and curious buddies. VLDB J. Int. J. Very Large Data Bases **20**(4), 541–566 (2011)

[OHA14]   O'Hagan, A.: No Place to Hide: Edward Snowden, the NSA and the Surveillance State by Glenn Greenwald. London Review of Books. Nicholas Spice. vol. 36, no. 18, pp. 11–12 (2014)

[PAV04]   Pavlov, E., Rosenschein, J.S., Topol, Z.: Supporting privacy in decentralized additive reputation systems. In: Trust Management, Lecture Notes in Computer Science, vol. 2995, pp. 108–119. Springer, Heidelberg (2004)

[PET]   Petrlic, R., Lutters, S., Sorge, C.: Privacy-preserving reputation management. In: Proceedings of the 29th Annual ACM Symposium on Applied Computing, New York, USA, pp. 1712–1718 (2014)

[RAV16]   Raval, S.: Decentralized Applications. O'Reilly Media (2016)

[ROF13]   Roffo, G., Segalin, C., Vinciarelli, A., Murino, V., Cristani, M.: Reading between the turns: statistical modeling for identity recognition and verification in chats. IEEE AVSS, pp. 99–104 (2013)

[SAN16]   Sanchez, D., et al.: Co-utile P2P ridesharing via decentralization and reputation management. Transp. Res. Part C Emerg. Technol. **73**, 147–166 (2016)

[SCH16]   Schaub, A., Bazin, R., Hasan, O., Brunie, L.: A trustless privacy-preserving reputation system. IFIP SEC - Privacy (2016)

[SIP12]   Sirivianos, M., Kim, K., Gan, J.W., Yang, X.: Assessing the veracity of identity assertions via osns. IEEE COMSNETS (2012)

[SOS16]   Soska, K., et al.: Beaver: a decentralized anonymous marketplace with secure reputation. IACR Cryptography ePrint Archive, 464 (2016)

[SWE02]   Sweeney, L.: k-anonymity: a model for protecting privacy. Int. J. Uncertainty Fuzziness Knowl. Based Syst. **10**(5), 557–570 (2002)

[THE14]   He, B., Chen, C., Su, Y., Sun, H.: A defence scheme against Identity Theft Attack based on multiple social networks. Expert Systems with Application. Elsevier (2014)

Knowledge, Information and Creativity Support Systems (KICSS)

Empirical Testing of a Technique for Assessing Prior Knowledge in the Field of Research Training

Thomas Köhler[1(✉)], Bahaaeldin Mohamed[2], Stefanie Seifert[3], and Thorsten Claus[4]

[1] Faculty of Education, TU Dresden, Dresden, Germany
thomas.koehler@tu-dresden.de
[2] Teacher Training College, Riyadh, Saudi Arabia
admin@phd-lab.com
[3] International University Institute, Zittau, Germany
seifert@ihi-zittau.de
[4] Faculty of Economics, TU Dresden, Dresden, Germany
thorsten.claus@tu-dresden.de

Abstract. As discussed in the research of Mohamed et al. (2013), learning and studying via a different context, including the inter-cultural knowledge of a teacher, may present a challenge when trying to account for students' prior knowledge. Enhancing their study, this paper addresses the question of whether or not an internal session comprising open discussion and brainstorming followed by asynchronous commenting and ending with self-reflection might help to assess students' prior knowledge. Data was collected in the context of a series of academic teachings partly supported by the METIS project (The METIS Project is funded by the national German Federal Ministry of Science and Research, cp.: http://www.ihi.zittau.de/cms/de/621/Forschung.) in two separate phases. Graduate students on master level attended the phase reported here in order to study research methods as a joint activity of Technische Universität Dresden, and Internationales Hochschulinstitut Zittau, both Germany. For the experimental group, the study found that there are no significant differences between the control and treatment group in terms of cognitive achievement and students' attitudes. From the teacher's perspective, however, this course is learning atmosphere challenged students' reactions. Design is suggested to be enhanced to allow a closer look on different types of prior knowledge in the field of research training.

Keywords: Prior knowledge · Academic achievement · Learning performance
Knowledge measurement · Test methodology · Research training
Post-graduate studies

1 Introduction

1.1 Students' Prior Knowledge: A Theoretical Framework

Mohamed et al. (2013) as well as Köhler and Mabed (2015) mention that the academic achievement of students is a top concern of educators and teachers, which ideally provides evidence for important aspects like future career opportunities and thus has a special meaning in the general, vocational and academic education fields. It is always necessary to

© Springer Nature Switzerland AG 2019
T. Theeramunkong et al. (Eds.): iSAI-NLP 2017, AISC 807, pp. 219–228, 2019.
https://doi.org/10.1007/978-3-319-94703-7_20

understand how previously acquired knowledge affects competency development. This is especially important for subject areas that are related to different domains and may be based upon differently acquired previous knowledge. Such a case is the knowledge about scientific working and research methodology, which is linked to very different academic disciplines.

Several studies have discussed students' pre-knowledge but do not especially address the issue of research methodology knowledge. However, these studies emphasize the importance of prior knowledge for reconstructing learning and understanding individuals, particularly learners who lack prior knowledge and could face increased difficulties and challenges during the learning process and the construction of learning. For example, specific prior knowledge may directly influence students' performance during the learning process (Alexander and Judy 1988). Prior knowledge may reflect an individual's personal explanation for events and actions that make sense to them but may or may not be in accordance with other explanations (Loughran 2010). In ninety-five percent of the literature, prior knowledge was found to have a positive effect on learning. Varying levels of prior knowledge can be supported when teachers take the likelihood of misunderstanding into consideration at the beginning of the learning process.

1.2 Types of Prior Knowledge: Cognitive, Psycho Motor and Emotional Components

Dochy (1994) reports that prior knowledge has to be classified into two main branches: domain-specific prior knowledge and general prior knowledge. In our research, domain-specific prior knowledge was considered to particularly influence students' performance (Hailikari et al. 2007). This knowledge not only pertains to the cognitive field, but also extends into learners' beliefs and attitudes that they bring with them into the classroom. In this context, prior knowledge represents what we have heard, what we have seen, what we have practiced, our complete emotional spectrum and attitude and, in some cases, what we have performed, our "psycho motor." These forms of knowledge can be later recalled in different ways as it is retrieved and applied to new and different situations (Loughran 2010). In this study, prior knowledge is not only considered as a cognitive component, but is also extended to moments of cognition, emotion, and, in some cases, psycho-motoric behavior.

During this research, the teacher facilitated a brainstorming session as part of a research method seminar as a means to further connect students with the problem/doubt they were experiencing and to bring students and teachers together so that they can understand, precisely define, and seek a solution to solve the problems challenges students were facing. Additionally, students should continuously comment and reflect on their learning during the seminar and, additionally, by documenting through a portfolio.

Due to the fact that the content of our seminar consisted of information about research methodology in the social sciences, this study is limited to dealing with only two types of knowledge: the cognitive domain, which reveals mental skills and knowledge, and the affective domain, which is connected with growth of feeling and attitude (Bloom and Engelhart 1969). The psychomotor domain is not included in this research, but would have been relevant for a study on research methodology in the natural sciences when research activities included certain lab or environmental practices (cf. El-Sabagh and Köhler 2010).

Over the course of the past few decades, many educational psychologists have identified a multitude of factors that might influence student achievement; in these studies great importance was placed on the influence of students' prior knowledge on learning and performance. Empirical evidence from educational literature highlights that student performance in an educational course is affected by his/her prior knowledge. Accordingly, the influence of the prior knowledge is essential for instructors (Hailikari et al. 2007). The responsibility of the teacher is to effectively use this prior knowledge and help students connect with new concepts and situations throughout the duration of the course. Additionally, studies also illustrate that domain-specific prior knowledge greatly influences student achievement when a teacher considers it when selecting a teaching method (Veselinovska et al. 2011). We therefore hypothesise **H1**: Students who discussed their domain-specific knowledge in depth with others have better cognitive performance than those without a structured discussion.

The affective domain of Bloom's taxonomy pertains to how to deal with emotional learning. Different definitions of the term attitude exist in the literature. Attitude includes five main components: emotion, goal, direction, strength, and consistency, each of which is considered either positive, neutral, or negative as attitude (Ashaari et al. 2010). Moreover it is shaped by the individual, their knowledge, and experiences and influences the individual's final attitude. Accordingly, for the purpose of this study, we assume the hypothesis **H2**: Students who discuss their domain-specific knowledge in depth with others have a more positive attitude towards learning than those without discussion.

2 Method

2.1 Research Design

The chosen research design has been presented first by Mohamed et al. (2013), based upon two main experiments (Table 1). The first was implemented for a course composed of local German students (N = 42) in the fall semester, while the second experiment focused on a course composed of international students (N = 12) during the summer semester. For Experiment A, the group was subdivided into a treatment and control group, while the experimental group B was approached as a treatment group only.

2.2 Subjects

Experiment A: This sample includes 42 German master's students who were studying in different disciplines across vocational education sciences, primarily healthcare, informatics, social pedagogy, philosophy, wood technology, machine technology, food and nutrition, chemistry, and metal technology departments at Technische Universität Dresden, Germany. Students participated in the same research methodology seminar in two different groups. The first group's seminar (Group A: treatment group, N = 18) took place between 4:40 pm and 8:00 pm every Tuesday evening for almost two months (from 26.11.2012 until 30.01.2013). The second group's seminar (Group B: Control group, N = 24) took place at a different time, between 1:00 pm and 4:20 pm every Monday. They started at the seminar 10.12.2012 and continued to have regular meetings from 07.01.2013 until 30.01.2013.

Table 1. Research design

Conditions	Experiment A	Experiment B
Instruments of measurements	• Multi-choice cognitive test (knowledge levels 1 & 2) • Likert-attitude scale	• Multiple-choice cognitive test (Knowledge levels 1 & 2[a]) • Essay test (Knowledge levels 3 & 4[b]) • Likert-attitude scale
Structure of measurement	2 groups (experimental & control group)	1 experimental group (pre-post)
Date of issue	26.11.12–30.01.13	15.04.13–15.07.13
Sample No.	N = 42	N = 12
Sample nationality	German students	International students
Treatment	Structured discussion/brainstorming session (Discuss, comment, and reflect)	

[a]1: Knowledge of facts, and 2: knowledge of meaning
[b]3: Integration of knowledge, and 4: application of knowledge

Experiment B: This treatment group consisted of 12 non-German students from Indonesia, Colombia, Brazil, Guatemala, Uzbekistan, Laos, Bolivian, Thailand, Mongolia and China. These students were enrolled in the "Vocational Education and Personal Capacity Building" programme and were also required to complete a research methodology seminar. The seminar started at 11:00 am each Wednesday and lasted almost 90 min for each week. For the purpose of the experiment, 7 seminars were conducted between 15.04.2013 and 15.07.2013. Due to the fact that this group was small, the decision was made to test it as one experimental group (pre-post) for higher-level knowledge, primarily knowledge of integration, and knowledge of application.

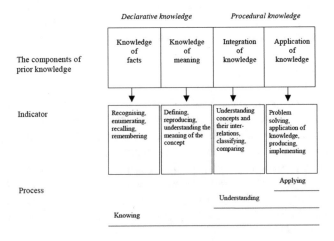

Fig. 1. The structure of knowledge levels

2.3   Measures and Procedures: Course Materials and Content Design

With the overall concept to develop students research competency the seminar, "Research methods: A critical analysis of empirical research" sought to assist students' acquisition of fundamental research concepts and to recognize terms that pertain to social science research methodologies. The seminar's goal was to not only foster students' understanding of how to quickly and productively find required scientific papers, but also to teach them how to read and understand and constantly evaluate the validity of scientific research. Therefore, students' tasks were planned to first facilitate students' recognition and comprehension of research methodology concept and then to apply this acquired knowledge in order to review a piece of research in depth. Accordingly, six full presentations (90 min), which had been completely developed by the teacher in order to guide students' knowledge, were used to serve as this seminar's learning material. The technology, PREZI (http://prezi.com/), was used to develop and deliver the seminar's content and was made available both in the classroom and online.

2.4   Development of Instruments for Measuring Student's Prior Knowledge

As described in the preceding publication (Mohamed et al. 2013) it was very important to define and recognize the students' background knowledge about this topic, consisting in Experiment A and Experiment B of (I) a cognitive multiple-choice test (30 items) to assess students' prior knowledge with test items based on knowledge levels 1 and 2 (knowledge of facts and knowledge of meaning) of prior knowledge. In parallel (II an attitude scale was applied, which consisted of two main parts (II_1) demographic data of the students and (II_2) their experience with reading and understanding scientific research, as a Likert type psychometric scale (Fig. 1).

Additionally for Experiment B an essay test was developed in order to cope with the student's high-level prior knowledge since the multiple choice test only measures the level of declarative knowledge.

3   Findings

3.1   Experimental Condition and Samples

Students in Group A (treatment group) recorded a high dropout rate (44.44%). This might be explained by the voluntary character of that activity in the context of the seminar and the relatively high effort for completing the trials in comparison to just a regular participation in the seminar. To cope with that situation we therefore excluded several data records from the data analysis as the participants had missed seminars, dropped out, failed to attend the required number of seminars, and/or failed to complete the final test. After excluding these records, we were left with a total of 30 usable records, with 8 students in the treatment group and a total of 20 students in the control group.

Concerning demographic information (Table 2), students in Groups A and B (control group) were quite heterogeneous. In terms of gender, Group A was 42.9% female and 57.1% male, while Group B consisted of 65.2% female and 34.8% male participants. In terms of disciplines and field backgrounds, Groups A and B include students from different fields (Group A: Chemistry 9.5%, Healthcare 28.6%, Wood Technology 9.5%, Informatics 9.5%, Metal Technology 9.5%, Philosophy 4.8%, and Social Pedagogy 4.8%; Group B: Construction Technology 17.4%, Chemistry 13%, Healthcare 17.4%, Wood Technology 8.7%, Food and Nutrition 13%, Metal Technology 13%, and Social Pedagogy 4.3%). Both groups were comprised of native Germans born in Germany (100%: N = 30). Although both groups took part in the seminar at different times of the day, they were generally offered the same number of meetings and the same learning condition regardless of whether the discussion/brainstorming session was included. Some additional meetings were arranged for the experimental group, but these meetings took place after the students had completed the post-test.

Table 2. Descriptive statistics (N = 30)

	Treatment	Control	Total
Males	6	7	13
Females	2	15	17
Age			24–33 years old

For the second section of descriptive data, students were asked to report their prior knowledge with regard to research methodology, particularly their ability to understand and review academic writing. This section was investigated through three main questions: the first related to whether or not students have prior knowledge of research methodology; the second asked if students had attended any related course; and the third asked students to self-evaluate their knowledge using a 5-point Likert-type scale, where one equaled low-level knowledge and 5 represented strong knowledge of the subject. In this descriptive section, "students' prior knowledge about research methodology," we illustrate how the results of these questions were used to determine "conflicts of interest" found among students who had completed similar courses before this investigation. The data depicted in Table 3 illustrates, however, that despite the fact that some students had completed similar seminars before, the students self-reported that they had limited experiences with the specifics of the seminar topic.

Table 3. Demographic data of sample, descriptive statistics (N = 12)

	Total	Completed records
Male	4	4
Female	8	5
Age	25–34 years old	

The third section of descriptive data pertaining to our attitude scale investigated self-assessment skills of student performance. This section consists of three main questions. The first asked students to estimate their own performance over the course of the last three semesters, where students had to select between 1, which meant "excellent", and 4, which denoted "pass" followed by the option "fail". The second question sought to gain insight into the students' capacity to study and how many hours per week students had to invest for learning. Students were also asked if they were completing full-time studies or if they were working in addition to their studies. The majority of students' performance was reported to be between levels 2 and 3 (n = 20, 66.66%), while 25 students (83.33%) reported that they studied intensively, more than 60 h a week. In addition, nearly the whole sample consisted of full-time students, which helped to provide more information about the students' schedules during the semester. Only one student reported that he was a part-time student (See Table 4).

Table 4. Descriptive statistics of Prior Knowledge (N = 12)

Prior knowledge		
	Total	Notes
Do you have any experience reviewing a scientific paper? (yes/no question)	Yes = 0	
	No = 9	
Have you completed any related courses about research methodology? (yes/no question)	Yes = 2	Students attended a similar seminar parallel to this study's seminar
	No = 7	
To what extent can you assess your understanding research methods? (Likert-type: 1, 2, 3, 4, 5)	Very low = 4	
	Low = 1	
	Middle = 4	

The final section in our descriptive data was used to uncover students' motivation vis-à-vis our seminar and the performance of the instructor. Due to the fact that this course is mandatory for all students, students were pushed to attend the seminar, regardless of our experiment. For post-evaluation of teacher performance, 73.33% of students reported that the teacher's performance ranked between the "average" and "good" levels of performance.

3.2 Student's Knowledge H1: Cognitive Domain

For hypothesis H1, the discussion session for the treatment group represented the independent variable. We therefore hypothesised that students who participated in discussion sessions performed better than those who had not. Table 6 presents the t-test results for our post-test. There was no significant difference between the treatment and control group in terms of performance (p = 0.33). As the main intervention, the discussion session had no influence on improving or worsening students' performance in our research methods seminar. Therefore, hypothesis H1 is not supported by the findings of this research. Due to the high number of drop-outs in the treatment group, a non-parametric Mann-Whitney test was performed as Chui et al. (2013) identify it as the most powerful test of the non-parametric tests and is a strong alternative to the t-test. Table 7 illustrates the non-significant differences between both groups and confirms that H1 cannot be supported by this study.

3.3   Student's Knowledge H1: Emotional Domain

This section presents students' emotional knowledge, i.e. their attitude towards learning in the seminar. Data collection with regard to the seminar's topic was conducted both before it began and after it had already ended. After analyzing data, authors calculated the mean of every category's item in order to determine students' overall attitude for each of the categories. Data show that no significant change occurred following our treatments and that only a small amount of change was noted in the category "prior knowledge of research methods," as students' attitude changed from "disagree" to "neutral." In general, no changes were recorded between the pre- and post-attitude scale concerning students' attitude towards our seminar. For the control group the situation was quite similar. Some attitudes changed in terms of students' soft skills and prior knowledge. Overall, no change was noted in students' attitude in the control group towards the seminar before and after attending class.

3.4   Experiment B

Experiment B began in the summer semester of 2013 by investigating 12 international students who were registered in the M.A. Program "Vocational Education and Personal Capacity Building." The students were tested as a single group by answering both the pre- and post-tests. 10 of the 12 students completed the post-test and attitude scale and, therefore, we excluded 2 records and were left with 10 usable records for the post-test and only 9 for the attitude scale (Table 3). Some students exhibited high-level self-motivation for gaining more information about research methods that would help them with the master's thesis they would be required to complete at the end of the program. The seminar took place each Wednesday in a conveniently located and well-equipped classroom.

Table 4 illustrates pertinent information about the students' prior knowledge of research methodology. Despite the fact that they were participating in an additional seminar about research methodology, they rated middle (n = 4) to very low (n = 4) knowledge about research methods. All students (n = 9) reported that they lacked a high-level of knowledge about the assessment skills required to evaluate academic research (Table 5).

Table 5. Descriptive statistics of students' performance frequencies (N = 9)

Students' performance frequencies		
	Conditions	N
Students' performance in the last 3 semesters	Level 1–2	2
	Level 2–3	6
	Level 3–4	1
How many hours of study per week?	4–20 h per week	
Full/part-time students	Full-time	

Table 6. Test of H1: students' discussion performance (pre t-test)

Performance category		Pre (t-test)	t (p)[a]
	Discussion treatment group	**Control group**	
	N=18 Mean (SD)	N=24 Mean (SD)	
Prior cognitive test	12.33 (3.39)	10.54 (3.13)	1.769 (0.084)

[a] P-value is two tailed

Table 7. Test of H1: students' discussion performance

Performance category		Post (t-test)	t (p)[a]
	Discussion treatment group	**Control group**	
	N=8 Mean (SD)	N=24 Mean (SD)	
Prior cognitive test	15.25 (2.92)	13.62 (6.22)	0.994 (0.329)
Performance category		Post (Mann-Whitney test)	t (p)[a]
	Discussion treatment group	**Control group**	
	N=8 Mean rank (sum)	N=24 Mean rank (sum)	
Prior cognitive test	17.38 (139)	16.21 (389)	89.000 (0.760)

[a] P-value is two tailed

4 Discussion and Conclusions

The findings of Experiment A identify that there was no significant difference between the control and treatment groups in terms of cognitive knowledge and non-significant changes in terms of emotional knowledge (students' attitude). These findings are inconsistent with the majority of previous research on students' prior knowledge, which emphasized the positive affect that reflection has on enhancing students' prior knowledge (Hailikari et al. 2007; Veselinovska 2011). For this experiment, three main factors may have strongly influenced the students' learning situation during the semester: The students' various/mixed domain specific knowledge, the teacher's knowledge background, and the learning environment/conditions.

Results of Experiment B demonstrate that there is obviously a lack of Prior Knowledge concerning research activity. This leads to a specific situation as this item doesn't seem to be distributed normally. In order to overcome this situation a next trial should be completed where further parameters can be measured as well.

Concerning the physical aspect of scientific work it should be discussed as well to extent the research design. Indeed a psychomotor domain needs to be included in further research taking into account the arguments of El-Sabagh and Köhler (2010). As well for the social sciences the nowadays widespread usage of (mobile/handheld) computing devices might be considered a psycho-motor activity that is relevant for a study on research methodology, not only in the natural sciences.

Given the presented results authors tried to investigate students' prior knowledge again with the same intervention (paper discussion session etc.) Yet for a successful research, it needs more – the handling of technical research environments and eventual real life ecologies. Such becomes observable only with the addition of a measurement tool. Hilikari (2007) reported that the different forms of knowledge should be measured by using different kinds of tasks. However, until now such procedural knowledge is hard to measure using a multiple-choice test alone. Thus, our results demand further attention. A selection of further samples might be appropriate, as this research has been completed with a small sample of a limited heritage only. To overcome this limitation authors already started a 2nd and 3rd experiment with further student groups. As well different procedures where applied, among them an additional essay test for higher levels of knowledge and a new type of experimental seminars with a linkage to practical research activity in the area research methodology training.

References

Alexander, P.A., Judy, J.E.: The interaction of domain-specific and strategic knowledge in academic performance. Rev. Educ. Res. **58**(4), 377–404 (1988)

Ashaari, N.S., Judi, H.M., Mohamed, H., Wook, T.M.T.: Student's attitude towards statistics course. Procedia Soc. Behav. Sci. **18**, 287–294 (2010)

Bloom, B.S., Engelhart, M.D., Furst, E.J., Hill, W.H., Krathwohl, D.R.: Taxonomy of Educational Objectives, The Classification of Educational Goals, Handbook I: Cognitive Domain. McKay, New York (1956), reprinted 1969

Chui, L., Martin, K., Pike, B.: A quasi-experimental assessment of interactive student response systems on student confidence, effort, and course performance. J. Acc. Ed. **31**, 17–30 (2013)

Dochy, F.: Prior knowledge and learning. In: Husen, T. Postlethwait, N. (eds.) International Encyclopedia of Education. pp. 4698–4702. Pergamon Press, London, New York (1994)

El-Sabagh, H., Köhler, T.: The impact of a web-based Virtual Lab on developing conceptual understanding and science process skills of primary school students. In: Gomez Chova, L., Marti Belenguer, D., Candel Torres, I. (eds.) ICERI2010 Proceedings CD (2010)

Hailikari, T., Nevgi, A., Lindblom-Ylänne, S.: Exploring alternative ways of assessing prior knowledge, its components and their relation to student achievement: a mathematics based case study. Stud. Educ. Eval. **33**, 320–337 (2007)

Köhler, T., Mabed, M.: Academic achievement in vocational secondary schools. Construction and validation of a test to assess potential learning performance; In: N.N. Advances in Intelligent Systems and Computing. Springer, New York (2015, forthcoming)

Loughran, J.: What Expert Teachers Do: Enhancing Professional Knowledge for Classroom Practice. Allen & Unwin, Crows Nest (2010)

Mohamed, B., Köhler, T., Seifert, S., Claus, T.: Discuss, comment and reflect: empirical testing of a technique for assessing prior knowledge. In: Threeramunkong, T., Yuizono, T., Skulimowski, A.M.J. (eds.) Proceedings of the 10th International Conference on Knowledge, Information and Creativity Support Systems. SIIT, Thammasat University Press (2013)

Veselinovska, S.S., Gudeva, L.K., Djokic, M.: The effect of teaching methods on cognitive achievement in biology studying. Procedia Soc. Behav. Sci. **15**, 2521–2527 (2011)

Problem Content Table Construction Based on Extracting Sym-Multi-Word-Co from Texts

Chaveevan Pechsiri[✉]

Department of Information Technology,
Dhurakij Pundit University, Bangkok, Thailand
itdpu@hotmail.com

Abstract. This research aims to construct a problem content table, particularly health-problem/symptom contents from downloaded health-care documents. The content table includes Disease Name, Symptom Concept, Symptom-Location Concept, and Sym-Multi-Word-Co Expression (which is a multi-word co-occurrence having a symptom concept on verb phrases). The research results benefit for a diagnosis system. The research has four problems; how to identify Sym-Multi-Word-Co from verb phrases, how to determine Sym-Multi-Word-Co boundaries after stemming words and eliminating stop words, how to solve Sym-Multi-Word-Co ambiguities, and how to derive symptom concepts and location concepts from Sym-Multi-Word-Co expressions with implicit-symptom-location occurrences. Therefore, we apply the symptom-verb-concept set to identify Sym-Multi-Word-Co and also to solve the Sym-Multi-Word-Co ambiguities. We also propose Bayesian Network to solve the Sym-Multi-Word-Co boundaries. We apply WordNet and MeSH to derive symptom concepts and implicit-symptom-location concepts. The research results provide the symptom content table with the high precision of the Sym-Multi-Word-Co extraction from the documents.

Keywords: Multi-word co-occurrence · Verb phrase · Symptom content

1 Introduction

The objective of this paper is to construct a problem content table, particularly health-problem contents as symptom contents extracted from health-care documents downloaded from the hospital's health-care web-board (http://haamor.com/, which is a non-government-organization website). The research emphasizes on extracting the symptom contents to construct the symptom content table. Each symptom-content consists of a symptom-term expression, a symptom concept, a symptom-location concept, and a disease name which is a document topic name. The symptom-term expression is based on the multi-word co-occurrence with the symptom concept from the web-board documents (where the 'multi-word co-occurrence' or 'multi-Word-Co' is the co-occurrence of two or possibly more of N-words; N = 2,3,.., *num* and *num* is the number of words per one sentence). The result of this research is the benefit for the automatic health-problem-diagnosis system and also the recommendation system. The symptom occurrence on a document can be expressed by either a noun/noun phrase, i.e. '*headache*' '*abdominal discomfort*', or a verb/verb phrase, i.e. '*vomit*' '*have tight chest*'. This research emphasize

© Springer Nature Switzerland AG 2019
T. Theeramunkong et al. (Eds.): iSAI-NLP 2017, AISC 807, pp. 229–243, 2019.
https://doi.org/10.1007/978-3-319-94703-7_21

only the verb phrase expression of each symptom because most of the symptom expressions in the documents, especially Thai health-care documents, are based on several consequences of events expressed by the verb phrases. Each verb phrase is based on the following linguistic pattern.

EDU → NP1   VP

VP → V1 | V1 NP2 | V1 Adv | V2 NP3 | V2 NP3 VP | V2 Adj

V1 → V_{strong} | Preverb V_{strong}

V2 → V_{weak} | Preverb V_{weak}

NP1→ Noun1 | Noun2 | Noun3

NP2 → Noun2 | Noun2 AdjectivePhrase

NP3→ Noun3 | Noun3 V1 | Noun3 AdjectivePhrase

Noun1→{'ผู้ป่วย/*patient*' 'โรค/*disease*'…}

Noun2→{'อวัยวะ/*organ*''บริเวณ/*area*' 'ศีรษะ/*head*''หน้าอก/*chest*''กระเพาะ/*stomach*' …}

Noun3→{'อาการ/*symptom*' 'แผล/*scar*' 'รอย/*mark*' 'ไข้/*fever*''ผื่น/*rash*''หนอง/*pus*''อุจจาระ/*stool*'…}

V_{strong}→{'คลื่นไส้/*nauseate*' 'อาเจียน/*vomit*' 'ปวด/*pain*' 'เจ็บ/*pain*' 'แน่น/*constrict*' 'คัน/*itchy*'…}

V_{weak} → {'เป็น/*be*' 'มี/*have*''รู้สึก/*feel*' }

Adv→ {'ยาก/*difficultly*' …. }   Adj → {'สี…/…*color*' 'เหลว/*watery*' ….}

Preverb → {'ไม่/*not*' … }

where EDU is an elementary discourse unit or a simple sentence [1].
NP1, NP2, and NP3 are noun phrases. VP is a verb phrase.
V_{strong} is a strong verb concept set with the symptom concept.
V_{weak} is a weak verb concept set which need more information to have the symptom concept.
Noun3 is a noun concept set with the symptom concept.
Adv is an adverb concept set with the symptom concept. Adj is the adjective concept set with the symptom concept. prep is a preposition.

Verb Phrase Example:

(a) "(เป็น/ *be*)/weak-verb (ผื่น/*rash*)/noun (แดง/*red*)/Adj (บน/*on*)/prep (หน้า/*face*)/noun"
      ("***To be red rashes on the face***")
(b) "(มี/*have*)/weak-verb (ผื่น/*rash*)/noun (ทั้ง/*all*)/det (ตัว/*body*)/noun"
      ("***To have rashes on all body***")
(c) "(มี/ *have*)/weak-verb (อาการ/*symptom*)/noun (คลื่นไส้/*nauseate*)/strong-verb "
      ("***To have a nauseated symptom***")
(d) "(รู้สึก/ *feel*)/serialverb (คลื่นไส้/*nauseate*)/strong-verb (ทั้ง/*all*)/det (วัน/*day*)/noun"
      ("***To feel nauseated all days***")
(e)   "(รู้สึก/*feel*)/serialverb (แน่น/***constrict***)/strong-verb (มาก/ ***badly***)/Adv (ที่/*at*)/prep (หน้าอก/***chest***)/noun (ของฉัน/***my***)/Adj "
      ("***To feel constricted badly on my chest***")

According to the (a)–(e) examples, each verb phrase expression contains the multi-word co-occurrence having the problem/symptom concept after stemming words and eliminating stop words as shown in the following expression named 'Problem-Multi-Word-Co' or 'Sym-Multi-Word-Co'.

Sym-Multi-Word-Co Expression:

Sym - Multi - Word - Co $= w_1 + w_2 + .. + w_{num}$ (where w_1 is $verbCo$ as a starting word of a Sym-Multi-Word-Co expression; $verbCo \in V_{weak} \cup V_{strong}; i = 2, 3, .., num$; $w_i \in$ Noun2 $\cup$ Noun3 $\cup V_{strong} \cup$ Adj $\cup$ Adv)

The Sym-Multi-Word-Co expressions from the (a)–(e) examples exist as one attribute of the symptom content table as shown in Table 1 which consists of four attributes; Sym-Multi-Word-Co Expression, Symptom Concept, Symptom-Location Concept, and Disease Name. The symptom concepts and the symptom-location concepts are referred to WordNet [2] and Mesh (https://www.nlm.nih.gov/mesh/) after translation from Thai to English by Lexitron (http://www.longdo.com/).

Moreover, previous researches have worked on event extraction from text [3–5]. In 2011, Fader et al. [3] applied syntactic and lexical constraints on binary relations expressed by verbs. Ando et al. [4] proposed methods for filtering harmful sentences based on multiple word co-occurrences. They compare harmless rate between two-word co-occurrence and three-word co-occurrence. And, Riaz and Girju [5] worked on a model for identifying causality in verb-noun pairs to encode cause or non-cause relation. Most of previous researches identify an event by two-word co-occurrence as verb-noun co-occurrence.

However, the research contains four main problems (see Sect. 3), the first problem is how to identify the verb phrases having Sym-Multi-Word-Co from the downloaded documents. The second problem is how to determine the Sym-Multi-Word-Co size/boundary after stemming words and eliminating stop words from the verb phrase (see Table 1). The third problem is the ambiguous Sym-Multi-Word-Co. And, the fourth problem is how to derive the symptom concepts and the symptom-location concepts from the extracted Sym-Multi-Word-Co with the implicit-symptom-location occurrences. Therefore, the research applies the symptom verb concept set, $V_{weak} \cup V_{strong}$, along with NP1 to identify the verb phrase having Sym-Multi-word co and also to solve the Sym-Multi-Word-Co ambiguity. The symptom expressions on the health-care documents mostly are based on the events expressed by verb phrases having some sequence words with conditional dependency, i.e. (เป็น/*To be* ผื่น/*rash*)➜ (แดง/*red*)" and "(เป็น/*To be* ผื่น/*rash* แดง/*red*)➜(หน้า/*face*|*skin*) . Thus, we propose using Bayesian Network (BN) [6] with *verbCo* and w_i features to solve the Sym-Multi-Word-Co boundary from the hospital web-board documents. We then integrate the word concepts of the following sets: V_{strong}, Noun2, Noun3, Adj and Adv for deriving the symptom concepts and the explicit/implicit symptom-location concepts from the extracted Sym-Multi-Word-Co expressions where all concepts are referred to WordNet [2] and MeSH, after translating from Thai to English by Lexitron. The extracted Sym-Multi-Word-Co expressions along with the disease names from the document topic names, and the derivation of both symptom concepts and symptom-location concepts from Multi-Word-Co expressions are organized into the symptom content table including the following attributes, Disease Name, Sym-Multi-Word-Co

Expression, Symptom Concept, and Symptom-Location Concept as shown in Table 1 which can be applied for the automatic problem diagnosis system.

The research consists of 5 sections. In Sect. 2, related works are summarized. The research problems of extracting the symptom contents to construct/form the symptom content table from the health-care documents are described in Sects. 3 and 4 is the research framework. And, we evaluate and conclude our proposed model in Sect. 5.

Table 1. Symptom content table extracted from health-care documents

Verb phrase example	Sym-Multi-Word-Co Expression	Symptom Concept	Symptom-Location Concept	Disease Name
(a)	'เป็น/**be** ผื่น/**rash** แดง/**red** หน้า/**face**'	To be red rush	face	Food Poisoning
(b)	'มี/**have** ผื่น/**rash** ตัว/**body**'	To have rush	body	Food Poisoning
(c)	'มี/**have** อาการ/**symptom** คลื่นไส้/**nauseate**'	To nauseate	stomach (from 'nauseate' by WordNet)	Food Poisoning
(d)	'รู้สึก/**feel** คลื่นไส้/**nauseate**'	To nauseate	stomach	Diarrhea
(e)	'รู้สึก/**feel** แน่น/**constrict** หน้าอก/**chest**'	To constrict	chest	Heart disease

2   Related Works

There are many research works on event expression, but only a few researches on identifying or determining an event by a word co-occurrence as in [3–5].

Fader et al. [3] introduced a syntactic constraint including an intuitive lexical constraint to identify the binary relations expressed by verb phrases (called relation phrases) for the Open Information Extraction system, REVERB. Their method can solve the incoherent and uninformative extractions. They implemented the syntactic constraints to REVERB for capturing relation phrases expressed by a verb-noun combination, including light verb constructions to match the POS tag pattern. The lexical constraint used to separate valid relation phrases from over specified relation phrases on the extraction as shown in the following.

"Faust made a deal with the devil." can be extracted as *"Faust, made a deal with, the devil"* instead of *"Faust, made, a deal"*.

The research results with more than 30% of REVERB's extractions are at precision 0.8.

Ando et al. [4] proposed using the positive (harmless) probability of each word co-occurrence from a certain sentence on Social Network Service for filtering negative (harmful) sentences or positive sentences. They defined a list of 'black words' and 'gray words'. If a sentence contains a gray word, the HarmlessRate value is calculated from the average positive probability of the two-word co-occurrence and three-word co-occurrence. If the HarmlessRate value of all word co-occurrence from a certain sentence is less than 0.3, the sentence is 'Harmful'. The precision of identifing and filtering the harmful sentences through three-word co-occurrence method exceeds 90% whereas the precision of the two-word co-occurrence is lower than 50%. However, their word co-occurrences used for filtering the harmful sentences are based on the surface forms.

Riaz and Girju [5] applied Integer Linear Programming to learn the causal relation from the annotated verb-noun pairs (from verb-noun phrase pairs) based on FrameNet, WordNet and the semantic class knowledge along with linguistic features. For example: "*People died in hurricane*" had the noun '*hurricane*' and the verb '*die*' being in the same event which resulted in annotating as 'Causal'. Whereas "*He presented further evidence in his presentation*" had the noun '*presentation*' and the verb '*present*' being annotated as 'NonCausal'. The result of this research achieves 14.74% and 41.9% F-scores for the basic supervised classifier and the knowledge of semantic classes of verbs respectively.

The previous researches [3, 5] works on verb phrases to express the events and also the relations where [3] applies the linguistic constraints to extract all words from verb phrases and [5] is based on two-word co-occurrence of 'verb-noun'. Whereas [4] works on filtering harmful sentences based on probabilities of two/three-word co-occurrences. However, our research concerns in determining the symptom event concept expressed by the Sym-Multi-Word-Co expression (N-word co-occurrence) on the verb phrase for constructing or forming the symptom content table.

3   Research Problems

There are four main problems of constructing the symptom content table based on extracting the Sym-Multi-Word-Co from the health-care documents; how to identify the verb phrase having a Sym-Multi-Word-Co expression, how to determine the Sym-Multi-Word-Co boundary on the verb phrase, how to solve the Sym-Multi-Word-Co ambiguity, and how to derive the symptom concepts and the symptom-location concepts from the extracted Sym-Multi-Word-Co expressions with the implicit-symptom-location occurrences.

3.1   How to Identify Verb Phrase Having Symptom Concept

According to the hospital's health-care web-boards, there are several verb phrases with/without the symptom concepts as shown in the following examples:

EDU1 "หลังจาก/*After* คนไข้/*a patient* (ทาน/*has had* อาหาร/*a meal* จำนวน *1000*แคลลอรี่ต่อวัน/*with 1000 cal. Per day*)/VP

EDU2 " [คนไข้/*a patient*] (รู้สึก/*feels* ไม่สบาย/*uncomfortable*)/VP"

EDU2 contains a verb phrase (VP) with the symptom concept whereas EDU1 having VP without the symptom concept

3.2   How to Determine Boundary of Sym-Multi-Word-Co Expression

There are the various sizes of the Sym-Multi-Word-Co expressions on the verb phrase examples, e.g. from (a)–(d) examples and (f)–(i) examples, which result in determining the Sym-Multi-Word-Co boundary on each EDU's verb phrase after stemming words and eliminating stop words.

(f) VP ="(รู้สึก/***feel***)/serial-verb (ปวดเมื่อย/***sore***)/strong-verb (กล้ามเนื้อ/ ***muscle***)/noun (ที่/***at***)/prep (หัวไหล่/ ***shoulder***)/noun"
("***To feel sore muscle at the shoulder***")
<u>Sym-Multi-Word-Co</u> = 'รู้สึก/***feel*** ปวดเมื่อย/***sore*** กล้ามเนื้อ/***muscle*** หัวไหล่/***shoulder***'

(g) VP = "(มี/***have***)/weak-verb   (ไข้/***fever***)/noun"
("***To have fever***")
<u>Sym-Multi-Word-Co</u> = 'มี/***have*** ไข้/***fever***'

(h) VP="(เป็น/***be***)/verb (เม็ด/***bumps***)/noun (พอง/***blister***)/noun (น้ำใส/***watery***)/Adj   (จำนวนมาก/***a lot***)/Adj "
("***To have a lot of watery-blister bumps***")
<u>Sym-Multi-Word-Co</u> = 'เป็น/***be*** เม็ด/***bumps*** พอง/***blister*** น้ำใส/***watery*** '

(i) VP = "(มี/***have***)/weak-verb (อาการ/***symptom***)/noun (อุจจาระ/***stools***)/noun   (เหลว/***watery***)/ Adj (หลาย/***several***)/Adj (วัน/ ***day***)/noun"
("***To have a symptom of watery stools within several days***")
<u>Sym-Multi-Word-Co</u> = 'มี/***have*** อาการ/***symptom*** อุจจาระ/***stools*** เหลว/***watery*** '

3.3   How to Solve Sym-Multi-Word-Co Expression Ambiguity

There are several Sym-Multi-Word-Co expressions contain symptom concepts on health care documents but a few of Sym-Multi-Word-Co expressions without having symptom concepts as shown in the following example.

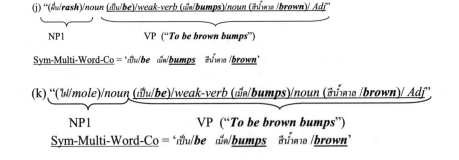

(j) "(ผื่น/***rash***)/noun (เป็น/***be***)/weak-verb (เม็ด/***bumps***)/noun (สีน้ำตาล /***brown***)/ Adj"

NP1   VP ("***To be brown bumps***")

<u>Sym-Multi-Word-Co</u> = 'เป็น/***be*** เม็ด/***bumps*** สีน้ำตาล /***brown***'

(k) "(ไฝ/***mole***)/noun (เป็น/***be***)/weak-verb (เม็ด/***bumps***)/noun (สีน้ำตาล /***brown***)/ Adj"

NP1   VP ("***To be brown bumps***")

<u>Sym-Multi-Word-Co</u> = 'เป็น/***be*** เม็ด/***bumps*** สีน้ำตาล /***brown***'

According to the above example, VP of (j) has the Sym-Multi-Word-Co expression with the symptom concepts whereas VP of (k) has the Sym-Multi-Word-Co expression with the property concept of NP1 or '*mole*'.

3.4 How to Derive Symptom Concept and Explicit/Implicit Symptom-Location Concept from Extracted Sym-Multi-Word-Co

The extracted Sym-Multi-Word-Co expression consists of the symptom concept expression and the explicit-symptom-location concept expression such as *'face' 'body' 'chest' 'shoulder'* and etc., as shown in the (a), (b), (e), and (f) examples

(a) 'เป็น/**be** ผื่น/**rash** แดง/**red** หน้า/**face**'   (b) 'มี/**have** ผื่น/**rash** ตัว/**body**'

(e) 'รู้สึก/**feel** แน่น/**constrict** หน้าอก/**chest**'

(f) 'รู้สึก/**feel** ปวดเมื่อย/**sore** กล้ามเนื้อ/**muscle** หัวไหล่/**shoulder**'

Whereas the other Sym-Multi-Word-Co expressions of the above examples have the implicit-symptom-location-concept occurrences, e.g. *'stomach'* in (c) and(d), *'body'* in (g), *'skin'* in (h), (j), and(k), and *'bowel'* in (i), as follow.

(c) 'มี/**have** อาการ/**symptom** คลื่นไส้/**nauseate**'   (d) 'รู้สึก/**feel** คลื่นไส้/**nauseate**'

(g) 'มี/**have** ไข้/**fever**'   (h) เป็น/**be** เม็ด/**bumps** พอง/**blister** น้ำใส/**watery**

(i) มี/**have** อาการ/**symptom** อุจจาระ/**stools** เหลว/**watery**

(j)-(k) เป็น/**be** เม็ด/**bumps** สีน้ำตาล /**brown**

According these four problems, the research applies *verbCo* (which is either v_{stromg} or v_{weak} + *info* where *info* is w_2; *info* ∈ Noun3; Noun3 exists in either NP3 or NP1) to identify the verb phrase having the Sym-Multi-Word-Co and also to solve the Sym-Multi-Word-Co ambiguity. We also apply Bayesian Network with *verbCo* and w_i features to learn the Sym-Multi-Word-Co size/boundary. Moreover, the word concepts of the following sets: V_{strong}, Noun2, Noun3, Adj, and Adv but not including the common word occurrences on the sign and symptom domain e.g. 'อาการ/*symptom*' 'รู้สึก/*feel*' etc., are used for providing the symptom concepts and the symptom-location concepts of the extracted Sym-Multi-Word-Co expressions where all of these concepts are referred to WordNet and MesH after translating from Thai to English by Lexitron. The implicit-symptom-location concept can then be derived from Sym-Multi-Word-Co by the word description on WordNet of the Noun3's word elements.

4 A Framework of Constructing Symptom Content Table from Extracted Sym-Multi-Word-Co

There are five steps in our framework. The first step is the corpus preparation step followed by the step of the Sym-Multi-Word-Co learning. Then, the Sym-Multi-Word-Co extraction step are operated and followed by the derivations of the symptom and location concepts and the attribute collection as shown in Fig. 1.

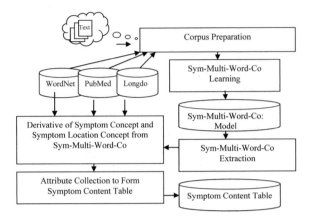

Fig. 1. System overview

Disease Topic : โรคเกี่ยวกับทางเดินอาหาร / **Gastrointestinal tract disease**

EDU1: ผู้ป่วยมีอาการจุกเสียดคออย่างมาก

ผู้ป่วย/*A patient* มี/*has* อาการ/*symptom* จุกเสียด/*be colic* อย่างมาก/*badly*

<EDU1>

(ผู้ป่วย /ncn)/NP1

(<**sym-multi-Word-CoExpression**   location= intestinal from WordNet of 'colic'>

  < **verbCo:** setType='weak-verb' ; concept='has/occur' boundary ='y'>มี</ **verbCo** >

  < w_2: setType='Noun3' ; concept= 'symptom' boundary ='y'>อาการ</w_2>

  < w_3: setType='strong-verb' ; concept=' be colic' boundary ='y'>จุกเสียด</w_3>

  < w_4: setType='Adv' ; concept= 'badly' boundary= 'n'>อย่างมาก</w_4>

</**sym-multi-Word-CoExpression** >)/VP </EDU1>

EDU2: [ผู้ป่วย]รู้สึกแน่นที่หน้าอกด้านขวาเป็นบางครั้ง

[ผู้ป่วย/*A patient*] รู้สึก/*feel* แน่น/*press against* ที่/*at* หน้าอก/*chest* ด้านขวา/*right side* เป็นบางครั้ง/*sometime*

<EDU2>

([(ผู้ป่วย/*A patient*)/ncn])/NP1

(<**sym-multi-Word-CoExpression**   location = chest from Noun2>

  < **verbCo:** setType='strong-verb' ; concept='oppress/press against' boundary ='y'>แน่น</**verbCo**>

  < w_2: setType='Noun2' ; concept= 'chest/organ' boundary= 'y'>หน้าอก</w_2>

  < w_3: setType='Adj' ; concept= 'right side' boundary= 'y'>ด้านขวา</w_3>

  < w_4: setType='Adv' ; concept= 'sometime' boundary= 'n'>เป็นบางครั้ง</w_4>

</**sym-multi-Word-CoExpression**>)/VP </EDU2>

The sym-multi-Word-CoExpression tag is the word boundary tag of each Sym-Multi-Word-Co expression including the symptom location property. The verbCo tag is the starting tag of the Sym-Multi-Word-Co expression. The w_i tag is the co-occurred word$_i$ tag where i=2,..,*num*. Both the verbCo tag and the w_i tag include the boundary property.

The [..] symbol means ellipsis (Zero Anaphora)

Fig. 2. Annotation of Sym-Multi-Word-Co

4.1   Corpus Preparation

This step is the preparation of corpus in the form of EDU from the medical-care-consulting documents on the hospital's web-board of the Non-Government-Organization (NGO) website. The step involves with using Thai word segmentation tools [7], including Named entity recognition [8]. After the word segmentation is achieved, EDU segmentation is then to be dealt with [9]. These annotated EDUs will be kept as an EDU corpus. This corpus contains 3000 EDUs of gastrointestinal tract diseases and childhood diseases and is separated into 2 parts; one part is 1000 EDUs of the gastrointestinal tract diseases and 1000 EDUs of the childhood diseases for learning the Sym-Multi-Word-Co expressions. And, the other part of 1000 EDUs from both the gastrointestinal tract diseases and the childhood diseases is for determining and extracting the Sym-Multi-Word-Co expressions. The symptom content table is then formed by the collection of the extracted Sym-Multi-Word-Co expressions including their disease names, symptom concepts and symptom-locations. In addition to this step of corpus preparation, the research semi-automatically annotates the Sym-Multi-Word-Co expressions and concepts of symptoms after the stop word removal as shown in Fig. 2. All word concepts of Sym-Multi-Word-Co expressions are referred to WordNet (http://word-net.princeton.edu/obtain) and MeSH(https://www.nlm.nih.gov/mesh/) after translating from Thai to English, by using Lexitron (the Thai-English dictionary) (http://lexitron.nectec.or.th/).

4.2   Sym-Multi-Word-Co Learning

BN [6, 10] represents the joint probability distribution by specifying a set of conditional independence assumptions (represented by a directed acyclic graph), together with sets of local conditional probabilities. Each variable in the joint space is represented by a node in BN. For each variable, two types of information are specified. First, the network arcs represent the assertion that the variable is conditionally independent of its non-descendants in the network given its immediate predecessors in the network. We say X is a descendant of Y if there is a directed path from Y to X. Second, a conditional probability table is given for each variable, describing the probability distribution for that variable given the values of its immediate predecessors. The joint probability for any desired assignment of values $\langle y_1, \ldots, y_n \rangle$ to the tuple network variables $\langle Y_1 \ldots Y_n \rangle$ can be computed by Eq. (1)

$$P(y_1 \ldots, y_n) = \prod_{i=1}^{n} P(y_i | Parents(Y_i)) \tag{1}$$

Where Y_0 is the parents of Y_1, and $Parents(Y_i)$ denotes the set of immediate predecessors of Y_i in the network. The values of $P(y_i | Parents(Y_i))$ are precisely the values stored in the conditional probability table associated with node Y_i. [6] also mentioned that the Bayesian structure could be constructed from the independence and dependence relationships from the data.

However, Eq. (1) is applied to the Sym-Multi-Word-Co boundary/size determination with $\langle Y_1 \ldots Y_n \rangle$ as consequence of words after the stop word removal, $\{w_1 \ldots w_n\}$, where Y_0 =Disease Topic from the document name. Each word, w_i (where i = 1..n), is a consequence word concept where w_1 is *verbCo; n = num*; i = 2,3,..,*num*; $w_i \in$ Noun2 ∪ Noun3 ∪ V_{strong} ∪ Adj ∪ Adv; and *verbCo* ∈ V_{weak} ∪ V_{strong}.

All annotated concepts of w_i are features for determining the conditional probabilities of a consequence words after the stop word removal as shown in Table 2.

Table 2. Show the sequence of w_i concepts appeared in the example documents of childhood diseases.

verbCo , w_1	P(w_1)	w_2	P(w_2|w_1)	w_3	P(w_3|w_2,w_1)	w_4	P(w_4|w_3,w_2,w_1)
·มี/*have*·		·อาการ/*symptom*·		อักเสบ/*inflame*	0.028571		
·มี/*have*·		·อาการ/*symptom*·		ท้องเสีย/*diarrhea*	0.014286		
·มี/*have*·		·อาการ/*symptom*·		หอบ/*dyspnea*·	0.014286		
·มี/*have*·		·อาการ/*symptom*·	0.071429	·ไอ/*cough*·	0.014286		
·มี/*have*·		·ไข้/*fever*·		·สูง/*high*·	0.028571		
·มี/*have*·		·ไข้/*fever*·	0.042857				
·มี/*have*·		·ผื่น/*rash*·		·แดง/*red*·	0.014286	·หน้า/*face*·	0.014286
·มี/*have*·		·ผื่น/*rash*·	0.028571				
·มี/*have*·	0.328571	.....	.....		.....		
·เป็น/*be*·		·ไข้/*fever*·	0.071429		0.071429		
·เป็น/*be*·		·ตุ่ม/*bump*·		·พอง/*blister*	0.014286	·น้ำ/*water*·	0.014286
·เป็น/*be*·		·ตุ่ม/*bump*·	0.028571				
·เป็น/*be*·		·ผื่น/*rash*·		·แดง/*red*·		·หน้าอก/*chest*·	0.007143
·เป็น/*be*·		·ผื่น/*rash*·	0.014286	·แดง/*red*·	0.014286		
·เป็น/*be*·	0.271429	.....	....		.....		
·ปวด/*pain*·		·ท้อง/*stomach*·	0.028571		0.02857		
·ปวด/*pain*·	0.042857	.....	....			....	
....	....	....	....	....	....		

From Table 2, It can concluded that the least probability of P(w_i|w_1,..w_{i-1}) is 0.007143 as the Sym-Multi-Word-Co Boundary threshold with the actual Sym-Multi-Word-Co Boundary threshold of 0.005 for determining the size or boundary of the Sym-Multi-Word-Co expression on the health care corpus, especially childhood diseases, as shown in the following rule (named the SymMultiWordCoBoundary rule)

$$\text{IF } P(w_i|w_{i-1}, ..w_2w_1) < MWC\_Threshold \quad \text{THEN}$$
$$SymMultiWordCoBoundary = \{w_1..w_i\}$$

where *MWC_Threshold is* the actual Sym-Multi-Word-Co Boundary threshold, and w_i = a consequence word concept after stemming words and the stop word removal.

4.3   Sym-Multi-Word-Co Extraction

The Sym-Multi-Word-Co extraction consists of two sub-steps: Identification of Starting Sym-Multi-Word-Co on a verb phrase and Determination of Sym-Multi-Word-Co Boundary.

4.3.1   Identification of Starting Sym-Multi-Word-Co

The research applies *verbCo* (which is either v_{stromg} or v_{weak} + *info/w_2* where $v_{stromg} \in V_{strong}$, $v_{weak} \in V_{weak}$, *info* $\in$ Noun3; Noun3 exists in either NP3 or NP1) to identify the starting Sym-Multi-Word-Co as the verb phrase having the symptom concept. Moreover, V_{weak}, V_{strong},Noun2, Noun3, Adj, and Adv are collected from the corpus behavior study on the annotated corpus.

Assume that each EDU is represented by (NP1 VP). Some verb phrases (VP) contain the Sym-Multi-Word-Co expressions.

```
MULTI_WORD_CO_BOUNDARY_DETERMIATON              /*by BN
 1   mWC←∅ ; verbCo ∈V_weak ∪ V_strong
        /* mWC is Sym-Multi-Word-Co expression.
        /* NP1word is a word occurrence on NP1.
        /* VPfirstWord is the first word of VP.
 2   If (NP1word∈Noun1∪Noun2∪Noun3)∧(VPfirstWord=verbCo )
 3   {   mWC← verbCo
 4       i=1
 5       MWC_Boundary ← P(verbCo)
 6       while MWC_Boundary>MWC_Threshold do
                /* MWC_Threshold  is the Sym-Multi-Word-Co
                   Boundary threshold from BN learning.
 7        { mWC ← mWC ∪ cw_i, i ← i + 1,
                /* cw_i is a consequence word right after
                   verbCo and after the stop word removal.
                /*cw_i∈ Noun2 ∪ Noun3 ∪ V_weak ∪ V_strong ∪ Adj ∪
                   Adv.
 8             MWC_Boundary ← P(cw_i|cw_{i-1}) * MWC_Boundary
                /*where cw_{i-1}∈Noun2∪Noun3∪V_weak∪ V_strong∪Adj∪Adv.
 9        } }
10   return  mWC
```

Fig. 3. The Sym-Multi-Word-Co boundary determination algorithm by BN learning

4.3.2   Determination of Sym-Multi-Word-Co Boundary

The Sym-Multi-Word-Co boundary is determined by using the SymMultiWordCo-Boundary rule with the conditional probability of each consequence word concept, w_i as shown in Fig. 3 of the Sym-Multi-Word-Co boundary determination after stemming words and the stop word removal.

Assume that mWC is the Sym-Multi-Word-Co expression object after stemming words and the stop word removal on the verb phrase. Loc is the body-part location set. SymCW is the symptom common word set.

Loc = {'*body*' '*Abdominal*' '*head*' '*chest*' '*skin*' '*stomach*' '*muscle*'...}

SymCW={'อาการ/*symptom*''รู้สึก/*feel*'}

Noun3={'อาการ/*symptom*' 'แผล/scar' 'รอย/*mark*' 'ไข้/*fever*''ผื่น/*rash*''หนอง/*pus*''อุจจาระ/*stool*'...}

Noun2={'อวัยวะ/*organ*''บริเวณ/*area*' 'ศีรษะ/*head*''หน้าอก/*chest*''กระเพาะ/*stomach*' ...}

SYMPTOMCONCEPT_LOCATION_ DETERMINATON

```
 1  {Symptom←∅ ;
 2   num1=NumberOfWord(mWC) /*count the number of words on mWC
 3   For i=1 to num1 do      /* derivation of symptom concept
 4     {If (mWC.word_i∉SymCW)∧(mWC.word_i∉Noun2)   then
 5         Symptom=Symptom + mWC.word_i ;
 6       i++;
 7     }
 8    Location=←∅; i=1; /*Derivation of
     explicitSymptomLocationConcept
 9   For i=1 to num1 do
10     {If (mWC.word_i∈ Noun2)   then
11     Location= Location + mWC.word_i;
12     i++;
13    } i=1 ;  j=1;    /*Derivation of implicitSymptomLocationConcept
14   while  Location=∅ ∧ i≤num1 do
15   { If (mWC.word_i∉SymCW) then
16      { descObject = WordNetDescriptionOfSynonym(mWC.word_i)
17        num2= NumberOfWord(descObject)
18        while  Location=∅ ∧ j≤num2 do
19          { If  descObject.word_j∈Loc then
20                Location= descObject.word_j;
21             j++;  }/*end-while Location=∅ ∧ j≤num2
22       } i++; }/*end-while Location=∅ ∧ i≤num1
23   return  Symptom, Location}
```

Fig. 4. The symptom-concept and the symptom-location determination algorithm from the extracted Sym-Multi-Word-Co

4.4   Derivations of Symptom Concept and Symptom-Location Concept

Each symptom content consists of the extracted Sym-Multi-Word-Co expression (mWC) from the previous step (Sym-Multi-Word-Co Extraction), the symptom concept, the symptom-location concept, and the disease name from the document topic name. The symptom concept can be derived from the extracted mWC object by having each word of the mWC object ($mWC.word_i$) being the element of $V_{strong} \cup V_{weak} \cup$ Noun3 $\cup V_{strong} \cup$ Adj $\cup$ Adv (but it is not the element of the symptom-common-word set, {'อาการ/symptom''รู้สึก/feel'}), as shown in Fig. 4. The explicit symptom-location concept can be derived from mWC by $mWC.word_i$ being the element of Noun2 set. The implicit symptom-location concept can be derived from mWC by $mWC.word_j$ being the element of Noun3 where the synonym description of $mWC.word_i$ by WordNet has one or more words related to the human-body/part-of-human-body concept as shown in Fig. 4.

4.5   Attribute Collection to Form Symptom Content Table

After all attribute data of Sym-Multi-Word-Co Expression, Symptom Concept, Symptom-Location Concept, and Disease Topic Name have been solved, the symptom contents are collected to form the symptom content table which can be arranged by sorting according to each application perspective. For example: the symptom content table is sorted by Sym-Multi-Word-Co to generate the distinct Sym-Multi-Word-Co expression with identification as Sym-Multi-Word-Co-ID along with its symptom concept, a list/set of symptom-location concepts, and a list/set of disease names. Therefore, the symptom content table sorted by Sym-Multi-Word-Co is suitable for applying to an automatic health-problem diagnosis through a question as "What does it mean when I feel constricted around my right chest?" in the question answering system.

5   Evaluation and Conclusion

The corpus used to evaluate the proposed method of constructing the symptom content table based on extracting the Sym-Multi-Word-Co expressions on verb phrases after stemming words and the stop word removal consists of 500 EDUs of the gastrointestinal tract diseases and 500 EDUs of the childhood diseases. This evaluating/testing corpus is collected from the hospital web-boards of the NGO website. The evaluation of the research method is based on the evaluation of the Sym-Multi-Word-Co extraction having all word concepts referred to WordNet and MeSH after using the Lexitron dictionary whilst the symptom concepts and the explicit/implicit symptom-location concepts are derived from the extracted Sym-Multi-Word-Co. In addition, the evaluation of the Sym-Multi-Word-Co extraction is based on precision and recall judged by three experts with max win voting as shown in Table 3.

The average precision of the Sym-Multi-Word-Co determination is 92% with the average recall of 57.5%. The reason of low recall is the anaphora problem, especially with Noun3. For example: there are some pronoun words, i.e.

'(อะไร/*something*)/*pronoun*' '(บางสิ่ง/*something*)/*pronoun*', appearing among the consequence words of some verb phrases with the symptom concept, which result in the low recall as shown in the following

(1) VP="(รู้สึก/***feel***)/*serialverb*   (มี/***have***)/*weak-verb*   (บางสิ่ง/***something***)/*pronoun*   (ข้างใน/***inside***)/*prep*
(จมูก/***nose***)/*noun* (ระหว่าง/***during***)/*prep* (เวลาเช้า/***morning***)/*noun*"
(***"feel to have something inside the nose during the morning"***)
(2) VP="(รู้สึก/***feel***)/*serialverb*   (มี/***have***)/*weak-verb*   (อะไร/***something***)/*pronoun*   (รัดแน่น/***tight***)/*adj*
(บน/***on***)/*prep*   (หน้าอก/***chest***)/*noun*"
("feel to have something tight on the chest")

Therefore, the anaphora problems have to be solved before applying the Sym-Multi-Word-Co determination. Moreover, some documents of the childhood diseases express the certain disease symptom by the verb phrase with too long explanation, as shown in the following example, which results in the lower precision of the Sym-Multi-Word-Co determination than the gastrointestinal tract disease one.

(3) VP="(มี/***have***)/*weak-verb* (อาการ/***symptom***)/*noun* (บวม/***swell***)/*strong-verb* (แดง/***redness***)/noun
(และ/***watery***)/*adj* (อักเสบ/***inflammation***)/*noun*   (ที่/***at***)/*prep* เปลือกตา/***eyelid***"

The example (3) yields the incorrectness of Sym-Multi-Word-Co as มี/***have*** อาการ/***symptom***   บวม/***swell*** แดง/***redness*** และ/***watery***. The error of the example (3) can be solved by applying the other machine learning techniques. Thus, the symptom/health-problem content table formed by determining the symptom/health-problem contents from the extracted Sym-Multi-Word-Co on the health-care documents by this research is very beneficial not only for patients to understand the disease symptoms but also for the automatic health-problem diagnosis system. Moreover, the proposed method of this research can also be applied to the other areas such as the industrial finance problems.

Table 3. The evaluation of the Sym-Multi-Word-Co extraction

Disease type	Correctness of Sym-Multi-Word-Co determination	
	Precision	Recall
Gastrointestinal tract diseases	93%	45%
Childhood diseases	91%	70%

Acknowledgement. This work has been supported by the Department of Information Technology, Dhurakij Pundit University, Thailand. Moreover, Onuma Moolwat, Achara, and Uraiwan Janviriyasopak have contributed greatly in this research.

References

1. Carlson, L., Marcu, D., Okurowski, M.E.: Building a discourse-tagged corpus in the framework of rhetorical structure theory. In: Current Directions in Discourse and Dialogue, pp. 85–112 (2003)
2. Miller, G.: WordNet: a lexical database. Commun. ACM **38**(11), 39–41 (1995)
3. Fader, A., Soderland, S., Etzioni, O.: Identifying relations for open information extraction. In: Proceedings of the 2011 Conference on Empirical Methods in Natural Language Processing, pp. 1535–1425 (2011)
4. Ando, S., Fujii, Y., Ito, T.: Filtering harmful sentences based on multiple word co-occurrence. In: IEEE/ACIS 9th International Conference on Computer and Information Science (ICIS) (2010)
5. Riaz, M., Girju, R.: Recognizing causality in verb-noun pairs via noun and verb semantics. In: Proceedings of the EACL 2014 Workshop on Computational Approaches to Causality in Language, pp. 48–57 (2014)
6. Mitchell, T.M.: Machine Learning. The McGraw-Hill Companies Inc. and MIT Press, Singapore (1997)
7. Sudprasert, S., Kawtrakul, A.: Thai word segmentation based on global and local unsupervised learning. In: NCSEC 2003 Proceeding (2003)
8. Chanlekha, H., Kawtrakul, A.: Thai named entity extraction by incorporating maximum entropy model with simple heuristic information. In: IJCNLP 2004 Proceedings (2004)
9. Chareonsuk, J., Sukvakree, T., Kawtrakul, A.: Elementary discourse unit segmentation for Thai using discourse cue and syntactic information. In: NCSEC 2005 Proceedings (2005)
10. Jensen, F.V.: Bayesian Networks and Decision Graphs. Springer, Secaucus (2001)

Unpredictable Disaster Evacuation Guide for Weak People by Real-Time Sensor Network, SNS and Maps

Taizo Miyachi[1]([⊠]), Gulbanu Buribayeva[1], Saiko Iga[2], and Takashi Furuhata[3]

[1] School of Information Science and Technology, Tokai University, Hiratsuka, Japan
miyachi@keyaki.cc.u-tokai.ac.jp
[2] Keio Research Institute of SFC, Fujisawa, Japan
[3] University of Utah, Salt Lake City, USA

Abstract. Tokyo residents should prepare predicted future potential great earthquakes similar to 2011 East Japan earthquake. We built an autonomous warning system with real-time sensor networks, SNS (twitter), disaster server, web server, and Google map server for signs of unpredictable disasters. We propose a new evacuation guide system based on the warning system, well designed for weak residents and visitors. We investigated on the effectiveness of collaborations, among real-time sensor network, SNS (twitter), evacuation maps with disaster knowledge, and evacuation wide area maps for unpredictable scale of disasters. We also discuss how the guide system with evacuation maps should be utilized by helpers in a district community for the evacuation of both weak people and new residents.

Keywords: Evacuation guide · Unpredictable disaster
Real-time sensor network system · Evacuation map · Weak people evacuation

1 Introduction

The Great East Japan Earthquakes 3.11 in 2011 made a serious damage in Japan. For example, Fukushima nuclear power plants had some nuclear leak accidents influenced that about 320,000 people needed to evacuate from their home towns. Moreover, the Great Kanto earthquake 1923 in capital Tokyo sacrificed 38,000 people died by a conflagration. Therefore, a powerful inland earthquake in the Tokyo metropolitan area within 50 years has been scientifically predicted in Japan. The effective warnings and modified warnings for the unpredictable scale of disasters and tsunamis were often too late for the residents to safely evacuate. The government and companies often needed a long time to confirm the right information and even sometimes concealed inconvenient facts for them. Weak residents need find signs of danger, avoid serious risk of potential damages and evacuate to a safer direction and spot. Therefore, acquiring real-time information in unpredictable disasters and accidents could be very helpful for safer evacuation especially weaker people.

© Springer Nature Switzerland AG 2019
T. Theeramunkong et al. (Eds.): iSAI-NLP 2017, AISC 807, pp. 244–253, 2019.
https://doi.org/10.1007/978-3-319-94703-7_22

We propose new effective evacuation guide with autonomous real-time sensor network, SNS (twitter), disaster server, web server, and Google map server. We also investigated on the effectiveness of collaborations (a) real-time sensor network and SNS (twitter), (b) evacuation maps with disaster knowledge and (a), (c) evacuation wide area maps for unpredictable scale of disasters, and (d) evacuation guide for especially weak people and new residents. We also discuss how the guide system with evacuation maps should be utilized by the helpers in a district community for the evacuation of especially both weak people and new residents.

2   Serious Problems by Unpredictable Scale of Disasters

A powerful inland earthquake in the Tokyo metropolitan area within 50 years has been scientifically predicted in Japan. The reason of the predicted serious damages in Tokyo by the historical Japan Great Kanto Earthquakes in 1923 are a combination of several different types of causes [1], such as crowded wooden houses, narrow roads, flexible ground of volcanic soil and river sediment, strong wind, fire tornado, and change of direction of wind.

Experts in an academic filed can only predict a reasonable damage based on the old records of some apparent phenomena in the field within several decades since the diaster. The other phenomena are called "unpredictable scale." Recently, crustal movements and global warming have produced new types of serious disasters that no expert could predict. Therefore, residents need to find the signs of such embedded unpredictable scale of damages and survive by themselves even before assists by local government.

Examples of unpredictable scale of changes after the great earthquake that increased serious damages are following below.

(1) Many building collapses, especially on soft fragile river sediment ground obstructed fire engines to enter stricken areas since such building collapses were unpredictable change for firemen. The wooden houses burned more and spread fires [1].

(2) Change of the wind direction between day and night expanded the fire stricken areas in the Japan Great Kanto earthquake in 1923. Then, it was tragedy fires spread to the north during the day time and spread to the south in the night after people evacuated to shelters in south areas.

Major problems by great earthquakes can be classified into next four categories.

(P1) No real-time evacuation guide system in dangerous changes

(a) Invisible fire outbreaks in restaurants and wooden houses, (b) Maps of the latest area of spread fires, (c) Change of safe places, (d) Change in evacuation obstacles and assist for resident's, such as building collapse, fallen concrete block walls, road collapse, and traffic jams, (e) Actual strength of horizontal shaking at a local spot.

(P2) No real-time evacuation guide system against "unpredictable scale of dangerous changes." (f) large conflagration area, (g) danger getting close to the first safe place, (h) other safer places outside of the home town, (i) time of wind direction change, and (j) additional rescue if needed.

(P3) Normalcy bias. About 70% of residents unfortunately did not evacuate in The Great East Japan Earthquake in 2011 although they knew the tsunami warning of the government in text and reading out by TV announcers [2, 3]. Most residents in Tokyo might have similar normalcy bias in the next future Great Tokyo earthquake since they have been safe for 92 years. They need to acquire the motivation for proper evacuation actions and the chance of reconsidering evacuation from conflagration.

(P4) Wind information near nuclear power plants. The direction and strength of wind both in Tokyo area and near a nuclear power plant are very important for residents to safely evacuate from disasters and serious accidents in nuclear plant.

(P5) Late warning of embedded risks for new residents, aged people, and weak people. Government and firemen, etc. usually started evacuation assist after the first serious damage happed. Weak people usually had serious damages at the first phenomena of disaster since they lived in dangerous spots and older people could not acquire the information of signs of serious damages.

3   The Feature of Autonomous Warning and Collaboration with Signs of Disasters

Major roles of our new warning system during earthquake are (R1) to (R4).

(R1)   Distribute information about the first shockwave and horizontal shaking of earthquake in the local areas.

(R2)   Reduce psychological bias (Normalcy Bias and Catastrophe Forgetting) [2].

(R3)   Make the latest evacuation map for both healthy people and weak people with effective routes and safer routes to safe places by using the Big Data of sensor data, such as P/S-wave of earthquake, fire, gas, etc. Residents could guess the risk and scale of damages and guide weak people to evacuate in safer route.

(R4)   Distribute real-time information of dangerous local spots to not only healthy people in safer spots but also residents who live in risky spots, weak people, and new residents that cannot know the history of attacked spots.

Our new warning system mainly consists of Autonomous Sensor Network system (ASN) [10], disaster database server "CloRob," [6, 8] Google-map server, SNS module, and emergency web server (see Fig. 1) [7]. The sensor network system consists of "Earthquake Sub-system (ES)" with P/S-wave sensors, and "Multiple Sensor's Pack sub-system (MSP). MSP works with several sensors such as GPS, flame, two gas sensors, and TFT display. Autonomous collaboration among ASN, ES, MSP, CloRob, twitter, and Google-map API service enables users to know both signs of dangers in warning maps and the latest safe shelters in evacuation maps in real-time. ES can obtain earthquake signals of magnitude 3 by P-wave sensor and S-wave sensor. ES simultaneously sends earthquake warnings with GPS position data both to the twitter server and to the MySQL database server "CloRob" by Arduino Ethernet shield. ES also sends "CloRob logo" in the tweet to the twitter server in order to certify the sensor data. CloRob generates maps from the table's data using Google map API functions [5] and sends to the web server. Here we have an event map where fire/gas places are shown with red/blue markers respectively. The second map for residence evacuation shows the latest "local safe places," corresponding to real-time changes by both disasters and accidents after the great earthquake. The maps are provided by the system web site (ex. http://ictedu.u-tokai.ac.jp/miyachi/maps.php). The URL is described in texts in the tweet as a link to CloRob. The Internet is usually robust and was also available even in case of the Great East Japan earthquake in 2011.

A user can acquire the maps on a screen of mobile terminals when s(he) only touches the URL in blue color. The terminals can be also used by mobile batteries at the power failure as travelers and young people usually use. Helpers in the district community could also guide both older people and new residents to evacuate with evacuation maps although parents of the new resident would be absent.

4 An Evacuation Warning and Guide for Unpredictable Disasters

Local governments distribute hazard maps for predictable scale of disasters and warnings in the web pages. Public organizations could not distribute sensor data in real-time. A public official often needed a long time period to make sure the danger since (s) he sometimes could not understand a dangerous sign of unpredictable scale of disasters. Residents should find a dangerous sign of change (e.g. wind direction change) and

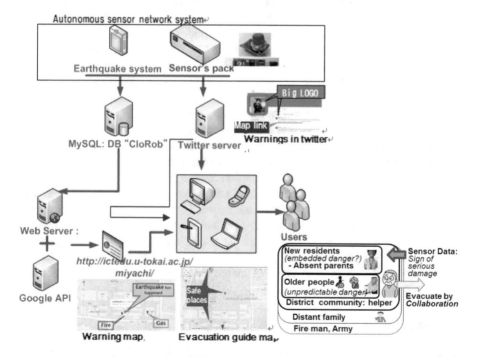

Fig. 1. Autonomous warning and guide system with real-time sensor data in "twitter," and evacuation maps guides both older people and new residents to evacuate by helpers collaboration in the district community.

signs of serious damages (e.g. fire in a restaurant) without getting help by public organization in order to take quicker and safer actions for effective evacuation from unpredictable scale of disasters in 24 h. We propose the new evacuation guide system utilizing the good aspects of real-time sensor networks, SNS (twitter), disaster server, web server, and Google map server for unpredictable disasters. We will discuss how to build effective collaborations in the utilization, such as (a) real-time sensor network and SNS (twitter), (b) evacuation maps with disaster knowledge and (a), (c) evacuation wide area maps for unpredictable scale of disasters and (a).

(a) **A combination of "autonomous sensor networks (ASN)" [4] and "SNS (twitter)"** distributes sensor data with GPS position to residents in real time with no delay. This combination consists of the fastest detectors of embedded dangers that human cannot discover and the fastest communication method between residents in the Internet. Residents that live in an embedded dangerous place (e.g. nearby fragile river sediment ground in Sumida district) can acquire a chance to find signs of serious damages (e.g. change of wind direction) in a variety of dangerous changes (e.g. collapsed wooden house) against unpredictable disasters and accidents. They could also start assist of weak people's early evacuation. ASN should have certification logos with GPS position data for the certification since information in SNS is not trusted.

(b) **Collaboration among ASN, SNS (twitter), CloRob, Google-map server, and emergency web server** delivers real-time awareness with GPS position in the emergency maps based on disaster knowledge in knowledge base CloRob [3]. Residents can easily understand real-time guides of both dangerous places and safe routes in the latest evacuation maps. They can also have chances to take quick actions and assist weak people by the maps in order to safely evacuate, reduce the serious damages, and cope with difficult problems caused by unpredictable disasters. The safe routes are shown by (1) the names of place, (2) an icon in a hazard map, (3) an arrow for evacuation route, and (4) an experience in evacuation training with the shelters corresponding to the kind of disaster. Residents could choose effective shelter for their situation.

(c) **Big-Data works for unpredictable scale of disasters and accidents.** A collection of real-time sensor data with GPS position could show the sign of the unpredictable situations for the residents. It would also show the expansion both by fire tornado and by the change of direction of wind between day time and night. The residents should find a safer place in the wider area than the nearest first shelter. They should also make a quick decision of long distance evacuation to the safer place, such as next town [9] and a big building in such situations. Weak people especially need safe shelters in the wider area since they cannot repeatedly evacuate dangerous route.

5 Experiment

Residents and their families could evacuate and reduce serious damages during the great disasters if our new system (real-time sensor data, discovery of the sign of danger, and the choice of an evacuation way) could be distributed in real time. We simulated how human subjects feel the danger of great earthquakes and how to safely evacuate by the evacuation guide system with real-time sensor data, distribution by twitter, evacuation maps with disaster knowledge.

Conditions. Suppose, (A) the subject lives in an earthquake area, and (B) the subject is a visitor. The earthquake warning has already been issued.

Devices. (1) Mobile phone and (2) TV.

Case(a) A subject that lives in an earthquake area (A) acquired the earthquake warning in a text-message on the screen of the mobile phone that automatically produced loud buzzer of the earthquake warning. The subject also acquires a twitter message that including real-time sensor data of the earthquake, fire, and gas, with GPS position that the autonomous sensor network systems detected and distributed to a subject's mobile phone. Twitter massages are shown with a large logo of "Banu-sensor-network" for the certification (See Fig. 1). Severe earthquake warnings were also shown on TV and an announcer of TV told the residents the warnings.

Case(b) A subject is a visitor (B). The subject acquired evacuation maps by a mobile terminal after the case (a). The maps are (i) Google map with the current position, (ii) Google map with icons that show real-time sensor data, such as earthquake, fire, and gas, according to GPS position, and (iii) Google map with certified safe places and dangerous places. The subject is going to evacuate by the warning (See Fig. 2).

Case(c1) A subject is a visitor (B). The subject acquired additional evacuation maps by a mobile terminal after the case (b). The maps show (iv) a large donut-shape of conflagration in unpredictable serious situations in wide Tokyo area, and (v) safe directions in (iv) (See Fig. 3).

Case(c2) A subject is a visitor. The subject acquired an additional evacuation map by a mobile terminal after the case (c1). The maps show (vi) an unpredictable serious situation with lots of fire icons and gas icons. This means that the first shelter has not already been safe. The subject should make a decision to evacuate to the outside of the town from the point of view to the safe place in the wide Tokyo area. (See Fig. 3).

(iv) Four conflagrations   (v) A large donut-shape of conflagration   (vi) Safe directions in a large donut-shape of conflagration

Fig. 2. Google Maps: A current position, Fire spots with 1st safe place, and Candidates of evacuation routes that avoids blind alleys

(i) A current position   *(ii) Fire spots with 1ˢᵗ safe place*   *(iii) Candidates of evacuation routes that avoids blind alleys*

Fig. 3. Google Maps: A current position, Fire spots with 1ˢᵗ safe place, and Candidates of evacuation routes that avoids blind alleys

Subjects: Group. 27 persons between 18 and 24 years old.

Operations.

(1) A subject watched and heard the explanations of real-time sensor data in twitter in Fig. 1.

(2) A subject watched and heard the explanations of real-time sensor data near Koenji station and evacuation guides in a mobile phone in Fig. 2.

(3) A subject watched and heard the explanations of serious damages in unpredictable disaster, rough evacuation guides in a mobile phone in Fig. 3.

Test.

Case(a)-1. Real-time sensor data in Twitter is instructive

Question1. "Are real-time sensor data with GPS position in twitter instructive?"

Answer1. yes/no/no answer: 24/1/2

Interview1-1. Do you trust information in twitter?

Answer1-1. Most subjects did not trust the information in twitter. They expected to refer the real-time sensor data and take safer actions than those without such data.

Interview1-2. What is necessary for you to trust the real-time sensor data in twitter?

Answer1-2. Big logo for certification of real-time sensor data.

Case(a)-2. Real-time sensor data in both evacuation map and twitter is instructive (Tokai Univ.)

Question2. "Can a resident evacuate to a safe place by evacuation map (Fig. 1.)?"

AnswerQ2. yes/no/no answer: 25/0/2

Interview2. Why a visitor can start the evacuation by the map (Fig. 1)?

AnswerI2. The subject could recognize the safe place since (s)he lived nearby there. (S)he knew the roads in the familiar place.

Case(b)-3. Real-time sensor data in both evacuation map and twitter is instructive (Suginami district)

Question3. "Can a visitor start evacuation to the first safe place by map Fig. 2 (ii)?"

Answer3. yes/no/no answer: 6/19/2

Interview3. Why a visitor could not start the evacuation by the map Fig. 2 (ii)?

AnswerI3.

(i)   The visitor wished not to take wrong actions. (S)he expected to follow other resident's evacuation since the residents knew the best way to evacuate from serious damages.

(ii) The visitor could not read the map of an unfamiliar place since (s)he could not have the idea of the roads and environments.

(iii) (S)he expected to evacuate in the correct direction. (S)he sometimes misrecognized the direction by 90°.

Case(b)-4. Real-time sensor data in both evacuation map with evacuation routes and tweets are instructive (Suginami district)

Question4. "Can a visitor start evacuation to the first safe place by map Fig. 2 (iii)?"

Answer4. yes/no: 24/3

Interview4. Why a visitor can start the evacuation by the map Fig. 2 (iii)?
AnswerI4.

(i)   The visitor could recognize the safe direction and concrete safe routes by the arrows in the map. Relationship between the shopping street, 1st safe place, and the arrows gave him/her the idea of the evacuation routes.

(ii) (S)he could find a safe route successfully far away from fires.

Case(c) Serious case in unpredictable scale of conflagration

Question5. "Can a visitor start evacuation to a safer area outside of donut-shaped dangerous flame by map Fig. 3 (v)?"

Answer5. yes/no/no answer: 4/14/9

Question6. "Can a visitor start evacuation to safer area outside of donut-shaped dangerous flame by map Fig. 3 (vi) with safe directions?"

Answer6. yes/no/no answer: 17/4/6

Intervie5. Why most visitors could not start the evacuation by the map Fig. 3 (v)?
AnswerI5.

(i)   They could not find a safe area. It was too large dangerous area to evacuate.

(ii) Some aged persons and I would be panic by unexpected serious situation.

Interview6. Why a visitor could start the evacuation by the map Fig. 3 (vi)?

AnswerI6. The visitor could recognize the approximate rough safe directions. (S)he strongly wishes to evacuate from the donut-shaped dangerous area. I could not help making a decision of the evacuation. The rough directions were valuable information.

Discussion. The combination of (a) real-time warning by twitter, (b) real-time sensor data on emergency maps, and (c) evacuation routes to safe place on evacuation maps were helpful for most subjects to start evacuation actions. We discuss the differences between subjects, following.

D1. Most subjects attempted to acquire real-time sensor data of accidents although they suspected such new information in the Internet. Some of them attempted to evacuate to the farest safe place from the dangerous spots even they had some doubt. Residents could avoid the late warning from the public association and fire men and select safer evacuation way before some serious damage would start.

D2. The arrow signs of the evacuation routes in the evacuation map were indispensable for many subjects who could not read the maps in unfamiliar towns since they could not autonomously started evacuation actions without the arrow signs. However, the new residents could know the safe evacuation route to a safe shelter even family or relatives were not there. They did not have to wait their family or relatives and they could voluntary evacuate themselves.

D3. More than half of the subjects that usually used Google map did not start evacuation by the maps since their friends and themselves sometimes had severe experiences of long time walking to reach the destination in some unfamiliar cities. Accuracy of the GPS reception in an area with either complex narrow roads or many tall buildings got worse than a few ten meters with more open field.

D4. Most subjects could immediately consider how to evacuate from the predictable serious disasters. However many subjects could not consider how to evacuate from the unpredictable scale of disaster. They only expected to follow the other resident's evacuation in such a situation. The residents could certify the safe route together by the evacuation route map and explain the safe route to new residents and weak people.

D5. Most subjects could make a decision of starting evacuation by the arrow signs of the approximately rough directions for safe areas in case of unpredicted scale of damages since they felt scare to survive.

D6. Helpers in a district community could understand both safe shelters and safe evacuation route by arrow signs in the evacuation maps. They could quickly explain such evacuation routes to both older people and new residents and evacuate with such weak people before severe damages would happen. They could also easily find a safer evacuation route with the map if the safe route in the map was already damaged.

6   Conclusion

We propose a new autonomous warning and evacuation guide system with emergency map and evacuation guide maps. There were some differences of abilities between subjects related to reading the maps and avoiding serious dangers. We could ensure that a combination of (a) real-time warning by twitter, (b) real-time sensor data on emergency maps, and (c) evacuation routes to safe place with some arrow signs on the evacuation maps were helpful for the most subjects to start evacuation themselves and collaboration with others.

References

1. Cabinet office, Government of Japan: Damage estimation and countermeasure of inland earthquake in the Tokyo metropolitan area (final report) (2013)
2. Kahneman, D., Tversky, A.: Prospect theory: an analysis of decisions under risk. Econometrica **47**(2), 313–327 (1979)
3. Miyachi, T., Buribayeva, G., Iga, S., Furuhata, T.: Evacuation assist from a sequence of disaster by robot with disaster mind. In: IWIN2013 (2013)

4. Miyachi, T., Buribayeva, G., Iga, S., Furuhata, T.: Discovery of precursors of serious damage by disaster context library with cross-field agents. eNOWS, pp. 118–123 (2014). ISBN 978-1-61208-329-2

5. Google Maps JavaScript API v3 documentation (2015). https://developers.google.com/maps/documentation/javascript/tutorial

6. Gulbanu, B., Miyachi, T.: Robots collaboration based on cloud robotics system for daily - emergency life. In: The 10th International Conference on Web Information Systems and Technologies (2014)

7. Gulbanu, B., Miyachi, T., et al.: An autonomous emergency warning system based on Cloud Servers and SNS. In: Proceedings of KES2015, vol. 60C, pp. 722–729 (2015)

8. Hu, G., Tay, W.P., Wen, Y.: Cloud Robotics: Architecture, Challenges and Applications. IEEE Network (2012)

9. Bureau of urban development Tokyo metropolitan government: Regional risk survey report on the earthquake (2008). http://www.toshiseibi.metro.tokyo.jp/bosai/chousa_6/download/houkoku.pdf

10. Moon, Y.B., Lee, J.Y., Park, S.J.: Sensor network node management and implementation, pp. 1321–1324 (2008). ISBN 978-89-5519-136-3

Author Index

© Springer Nature Switzerland AG 2019
T. Theeramunkong et al. (Eds.): iSAI-NLP 2017, AISC 807, pp. 255–256, 2019.
https://doi.org/10.1007/978-3-319-94703-7

Printed in the United States
By Bookmasters